An Introduction
to Woodland Ecology

An Introduction to Woodland Ecology

John Cousens

Senior Lecturer in Ecology
Department of Forestry and Natural Resources
University of Edinburgh

Illustrated by Tim Smith

Oliver & Boyd . Edinburgh

574·5264 COO (039735/75)

In the same series:
Mills: An Introduction to Freshwater Ecology
Gimingham: An Introduction to Heathland Ecology
FitzPatrick: An Introduction to Soil Science

OLIVER & BOYD
Croythorn House
23 Ravelston Terrace
Edinburgh EH4 3TJ
A Division of Longman Group Limited

ISBN 0 05 002775 1 (paperback)
     0 05 002869 3 (hardback)

First published 1974

Printed in Great Britain by Bell & Bain Ltd., Thornliebank, Glasgow

Contents

Introduction   1

1 Woodlands as ecosystems   5
2 How do woodlands change with time?   18
3 Succession and climax   26
4 The history of British woodland   38
5 Populations in the ecosystem   49
6 Assessing the relative importance of species populations
    I Primary producers   59
7 Assessing the relative importance of species populations
    II The decomposers   74
8 Assessing the relative importance of species populations
    III Herbivores, predators and parasites   87
9 Predictive models   99
10 Woodland types   107
Appendix I   What are ecosystems?   121
Appendix II   What to do and where to do it   125
References   139
Glossary   142
Index   147

Acknowledgements

I should like to say thank you to students and colleagues in the Department of Forestry of the University of Edinburgh who have assisted me greatly.

My acknowledgements are also due to the following for kindly allowing me to reproduce some of this material: J. A. R. Anderson (Fig. 3.3); Edward Arnold Ltd. (Figs. 7.2, 7.3); P. Bannister (Table 2.1); Cambridge University Press (Table 4.1); A. Carlisle, A. M. F. Brown, E. J. White (Fig. 5.5); Controller of Her Majesty's Stationery Office (Table 9.1); English Universities Press (Fig. 4.2); C. G. Hewitt (Fig. 8.2); E. W. Jones (Fig. 3.6); H. Klomp (Table 9.2 and Fig. 9.3); P. Lebrun (Table 7.1); O. W. Richards (Fig. 5.1); W. A. Thomas (Fig. 7.1); A. Tittensor (Table 8.1); J. Wyatt-Smith (Fig. 3.5); P. J. Zinke (Fig. 1.2).

Edinburgh, 1974   JOHN COUSENS

Introduction

Woodlands do not make up as much of the landscape in Britain as they do in much of Europe and N. America. Nevertheless, they form a major element in our vegetation and it may at first seem surprising that ecologists in Britain have paid so little attention to them during the last forty years. Glance through an ecology textbook such as that of Kershaw (1964) and you will find that virtually all the examples used which concern woodland are taken from N. American literature. British ecologists have tended to select other vegetation types for the testing of hypotheses about vegetation processes. There are technical difficulties in describing woodland or locating truly representative examples of it—the 'whole' is not visible at one time—there is such an enormous range in size between component species. Then Britain's woodlands are at the most 'semi-natural', as Tansley put it—meaning that trees will have been planted or removed or intensive grazing will have changed the character of the groundflora. Ecological interpretation is more straightforward if evidence of long past events does not remain to complicate the issues. Of the large areas of woodland established in the last few decades most are comprised of exotic species and the processes involved in the re-creation of the woodland environment are not yet complete.

A renewal of interest in woodland ecology came about in 1949 with the formation of the Nature Conservancy and the establishment of Nature Reserves, many of them in woodland. The formation of a reserve with a local warden in charge has often amounted to a significant change in environment, particularly in the more vigorous exclusion of domestic grazing animals. Changes in the character of the reserves began to be noticed. To maintain the original position active management was increasingly required. Research could be and was, initiated into these problems, but action was often needed before the results were available. For woodlands there was a large body of management knowledge accumulated by the Forestry

1

Commission and private forestry organisations but very little of this knowledge seemed to be directly applicable. The objectives of the Nature Conservancy were quite different and money available for management was on a smaller scale. How does one replace (foresters would say regenerate) an over-mature wood in such a way that there is minimal disturbance and maximal continuity? What are the features of a particular woodland that make it the preferred environment for a rare species of bird or moth? Questions like these underlined the need to know how woodland functioned in all its aspects. The scene was set for a new approach to woodland ecology—for the study of woodlands as ecosystems. Readers unfamiliar with the ecosystem concept should study Appendix I before embarking on the first chapter.

At the ecosystem level it is mandatory to consider plants and animals together, for, in theory at least, every species could have some interaction with every or any other species. This is what makes the ecosystem so complex and intimidates many a potential investigator. The extent to which we can unravel this complexity with useful effect depends on how the various ecosystem functions or processes are shared among the species present. When a few species dominate each process then simplification of the system in terms of these dominant species alone may be profitable. If the admittedly crude model we construct yields predictions that even approximate to reality then we have moved one tiny step forward. If the model can be applied to other ecosystems, then we can begin to generalise. As yet, the generalisations we can make are so broad that they may seem more 'common sense' than inductive reasoning of any great merit. Ecosystem ecology is beginning to evolve its own jargon, and at this stage some may rightly complain that the jargon makes essentially simple ideas more difficult to understand. Thus 'the primary production of an ecosystem sets a limit to production at higher trophic levels' is saying that the number of animals, parasites and decay organisms are ultimately dependent on the amount of food provided by green plants.

Today ecology is a household word. The reasons stem from the tremendous upsurge in world population and the simultaneous intensification of urbanization, technicological development and exploitation of natural resources. Pollutants are being released in ever increasing quantity with effects on the residual countryside and waters that we have seldom been able to predict. Define 'pollutant' as any substance present in such quantity that it prevents the normal functioning of an ecosystem and it becomes clear why ecosystem ecology must hold the key to this kind of problem. This is not to claim that ecosystem ecology should replace animal ecology or plant ecology. After all, its essence is the understanding of ecosystem processes and the identification of species making important contributions to them. To this end quantitative data are required, data of the kind that animal population ecologists and plant production ecologists are publishing every month. But this data is very unevenly spread. We know an enormous amount about the relatively small number of plants and animals that have been economically important to Man and a further few species that have been popular for experimental work, but very little about the others. All too often the ecologist is held up because there is little or no relevant information about one or other of the species of major importance in the ecosystem he is studying. The recent International Biological Programme represented a major step forward in this field of ecology. All over the world groups of ecologists were brought together to study major ecosystem types. What is to happen to them now that IBP is coming to a close? Will they be dispersed again and will we revert to a situation where research money is allocated only when a problem ecosystem has shown changes which may be irreversible at acceptable cost levels? Hopefully, in the future, we will be able to anticipate and prevent ecosystem breakdown, but, first, we need quantitative information about the critical pathways in ecosystem metabolism and the species activating them.

These are some of the reasons for adopting a framework of ecosystem ecology for this introduction to woodland ecology. If, after reading this book, just a few more people accept the challenge that the greater complexity of the woodland environment presents, if they begin to monitor changes taking place in their local woodlands, then something worthwhile will have been accomplished.

As a nation we are just beginning to appreciate that the quality of life is worth maintaining. The woodlands of Britain have a large part to play. They are not valuable simply as sources of timber and pulp-wood. They enhance the beauty of the landscape. They provide recreational facilities and quiet sanctuary. They are the homes of myriads of wild plants and animals. But they are also living entities. They cannot be preserved as they are. They inevitably change and they must be managed well if they are to continue to contribute to the quality of life. And we cannot manage them to conserve their most desirable features unless we know how woodland ecosystems function.

A feature of this series on ecology is the inclusion of suggestions for practical work—not mere exercises, but work that may add to our store of knowledge and help monitor the changes taking place in our woods. The usefulness of any ecological work is greatly increased if:

(*a*) other workers can come back to your area if they wish

Rule 1. Describe the exact location of the wood and the experimental or sample areas within it.

(*b*) comparisons are readily made with work in other areas

Rule 2. Describe the wood in its existing condition, the site and its major tree components.

Rule 3. Give details of the techniques you used and make some kind of assessment of likely precision or representativeness of your data.

Rule 4. List any unusual factors known to be operating.

Rule 5. List any past event that you have reason to believe may have contributed to the present situation.

In all ecological work a compromise has to be reached between the intensity of work that the problem seems to demand and the intensity that is practicable with the resources available. These resources are time, equipment and man-power with varying skills. In planning the suggestions for work that follow most of the chapters the assumption has been made that only relatively inexpensive equipment will be available and that the man-power will be intelligent but inexperienced. It follows that instruction in techniques must be thorough and supervision adequate if useful results are to be obtained.

1 Woodlands as ecosystems

Woodlands are unique among ecosystems in the size and longevity of their main primary producers, the tree species. Even the comparatively short-lived birches live longer than Man and reach over 20 metres in height. A long-lived species like the oak lasts several hundred years and in good localities may reach 30 metres. Exceptional individuals may live very much longer. The Winfarthing Oak in Norfolk was called an old tree at the time of the Norman conquest and Grigor estimated in 1841 that it could then be some 1600 years old. Remember, however, that many estimates of the ages of very old trees are no more than guesses, because the centre of the stem has rotted away and the annual rings are no longer there to count. The same problem arises in the wet tropics where growth-rings, if they are present, are produced irregularly and have no simple correspondence with age. The bristlecone pine (*Pinus aristata*) in the mountains of Arizona has individuals up to 4600 years old with growth-rings right to the centre. The mountain ash (*Eucalyptus regnans*) of Australia rises to over 122 metres and the giant redwood (*Sequoiadendrom giganteum*) can measure 12 metres round the base of its stem.

It has been estimated that the largest redwoods weigh over 500 tonnes and there are several trees to the hectare! As this represents an accumulation over more than 3000 years the annual average growth-rate is not so startling—at less than two tonnes per hectare per year. It is not unusual in many parts of the world for woodlands to accumulate biomass at the rate of over $2\frac{1}{2}$ tonnes per hectare each year over long periods. The faster-growing Scots pine woods in Britain achieve an average annual rate of more than 5 tonnes per hectare per year over the first 50 years of their life.

Most of this biomass accumulation goes into the massive framework of the ecosystem—into the support tissues of roots, stem and branches. There it is locked away for the greater part of the life-span of the tree.

The term *biomass* is at times confusing. It refers to the matter in living organisms, all of it, not merely the living tissues. In a large tree all but the outer, more recently laid down rings of wood in stem branch and root consist entirely of dead cells. Thus it happens that in any organism with a high proportion of dead skeletal material in its make-up, biomass is not, as its name certainly suggests, a good indicator of functional activity or maintenance requirements.

The support tissue of trees derives its strength from cellulose fibres embedded in a plastic matrix of substances known as 'lignins'. We do not yet know the molecular structure of lignins—but we know that they are not all identical. Some 300 million years ago lignins were organic substances being produced in quantity for the first time as woody plants began to dominate terrestial vegetation. There must be something very odd about the molecular structure of lignins, because very few saprophytes have managed to adapt themselves to exploit it—mainly the most advanced group of the Fungi, the Basidiomycetes. However, even basidiomycetes do not completely degrade lignins, tending to leave, in the litter and the soil, humic acid residues which are themselves slow to degrade in cool climates.

TABLE 1.1

Biomass components in a 40-year-old pine plantation
tonnes dry weight per hectare (assessed on ¼ ha)
Dominant, sub-dominant and suppressed tree classes.*

Component	Dominants		Sub-Dominants		Suppressed		Crop Totals	
	tonnes	%	tonnes	%	tonnes	%	tonnes	%
1. Cones	0·4	0·8	—	—	—	—	0·4	0·6
2. Needles	4·4	9·1	1·0	7·6	0·07	7·4	5·5	8·8
3. Branch wood	6·5	13·4	2·6	20·1	0·08	8·4	9·2	14·7
4. Branch bark	0·9	1·9	0·4	2·9	0·01	1·2	1·3	2·1
5. Stem wood	32·6	67·3	7·8	61·7	0·66	73·1	41·1	66·3
6. Stem bark	3·6	7·4	0·9	7·4	0·09	10·0	4·6	7·4
TOTALS	48·4	100	12·7	100	0·91	100	62·0	100
3+5 Support Tissue	39·1	80·7	10·4	81·8	0·74	81·5	50·3	81·0
No. trees per ha	760		624		200		1584	
Mean tree biomass (kg)	64		20		4·5		39	

* see text pp. 10–11.

Thus when twigs and branches fall to the ground they are only slowly decayed. With dead leaves, spent infructescences, bracts and bark flakes they constitute the 'litter', as it is called, which blankets the soil of most woodland. In boreal coniferous woodland the litter layer is often thick. The long cold winters inhibit decomposition to a greater extent than production and the litter layer thickness is the clue to this imbalance. In humid tropical climates where decomposition goes on at a high rate throughout the year there is no litter accumulation—the forest floor in tropical rain forest is almost bare. Even in dry tropical climates there is only periodic litter accumulation. The dominant organisms responsible for the breakdown of woody tissues in the tropics are undoubtedly the termites which comminute the stems of even hard-wooded trees in an incredibly short time.

Animals do not normally have the ability to digest celluloses let alone lignins (see Chapter 5). Termites can use wood as an energy source through the intermediaries of symbiotic gut protozoa which themselves have internal symbiotic bacteria. The D.O.M. (dead organic matter) of litter is broken down into smaller particles which become incorporated as 'humus' in the upper layers of the soil proper (the 'mineral' soil). In temperate zones where decomposition is generally less efficient and biochemical change limited, humus is black or dark brown: in the tropics it is colourless.

Each tree in woodland changes the environmental space into which it grows. The air space around it becomes more humid as water vapour is released in transpiration, the ground beneath becomes more densely shaded as the crown develops and in the soil the root-system causes local depletion of soil moisture and nutrients. Gaps between the trees allow light to reach the ground directly.

The temperature regime at ground level in a gap is a simple function of the size of the gap and the height of the canopy in which the gap occurs. Maximum surface temperature at the centre of the gap reflects the proportion of the day during which direct irradiation can take place. See Fig. 1.1 overleaf.

Table 1.2 summarizes the results of two assessments in successive years at a number of stations in a pine wood in Scotland. Both were made in October when the sun is fairly low in the sky, nevertheless they show clearly that the proportion of light reaching the ground increases when light in the open is reduced by an overcast sky. In interpreting data of this kind it must always be remembered that any vegetation type may thrive in full light for a limited time: the minimal percentages observed are indicative of the relative degree of shade tolerance of the groundflora species concerned.

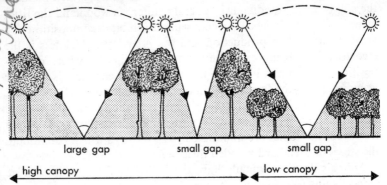

Fig. 1.1. Interaction of gap size and canopy height in determining the proportion of the day during which direct radiation reaches the ground.

TABLE 1.2

Light intensity at ground level in a wood as a percentage of light in the open—related to the type of ground vegetation.

Ground vegetation	Light intensity % (sunny day)	Light intensity % (overcast day)
1. None (bare litter)	4	9
2. Thin growth of moss	5	13
3. Full cover of moss	—	23
4. Grass cover (*Deschampsia flexuosa*)	9	17–50
5. Heather (*Calluna vulgaris*)	10–16	31–52

This is just one way in which the trees contribute to the areal pattern of the groundflora in woodland. They also intercept and redistribute rainfall and other types of precipitation, creating a pattern of varying moisture regimes. Figure 1.2 shows that on a neutral to slightly alkaline soil the soil becomes most acid near the tree trunk, while for reasons possibly related to the direction of the prevailing wind and its effect on litter fall the percentage nitrogen in the soil is highest to the NNW. Unless the soil and topography are regular and the canopy uniformly dense the groundflora pattern will be an irregular patchwork.

The vertical organisation of woodland structure is much more readily described and classified in terms of characteristic strata.

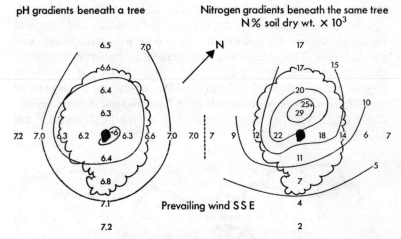

Fig. 1.2. Patterns in soil characteristics created by interaction between a tree and its environment (from: Zinke, 1962).

Strata that may be identified are:

1. the mainstorey which, as its name suggests, provides the general canopy. Occasionally there are isolated individual trees with crowns above the general canopy. Thus in tropical rain forest there is usually such an emergent stratum and an analagous situation exists in 'coppice-with-standards' woodland, the coppice providing the mainstorey.

2. the understorey composed of tree species too large to be called shrubs (over ten metres) and yet not reaching the mainstorey level.

3. the shrub layer of woody species less than 10 m and over 1 m tall.

4. the dwarf shrub layer of woody species less than 1 m tall.

5. the tall herb layer which may sometimes exceed 1 m.

6. the herb layer or field layer.

7. the ground layer, usually of mosses and liverworts.

Not all these strata will be present in one place. Indeed if the mainstorey forms a very dense canopy it may be the only stratum present. If the shrub layer is well developed the herb layer is likely to be

weakly developed and the canopy above rather open. An understorey
is a normal component of moist tropical forests but rather rare in
temperate climates except where underplanting has been done (e.g.
western hemlock, *Tsuga heterophylla*, under larch, *Larix* spp.).
Species of higher strata will appear in lower strata as seedlings (and
saplings). Figure 1.3 shows an oak wood with mainstorey and
shrub layers. The groundflora represented is rather more varied
than would occur in one small area. Bramble (*Rubus fructicosus*
agg.) is a dwarf shrub; bracken (*Pteridium aquilinum*) a tall herb or
'forb' (a term invented to include both ferns and angiosperm herbs);
dog's mercury (*Mercurialis perennis*) is on the borderline for tall
herb status.

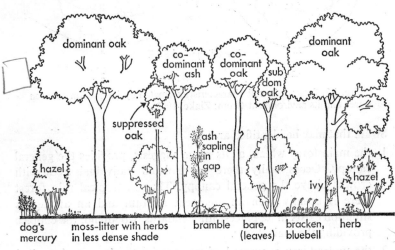

Fig. 1.3. Profile diagram of an oak wood showing strata— groundflora
not to scale.

Woodland plants may also be classified according to the way they
function, into categories called **synusiae**. The heterotrophic synusia
includes the parasites and hemi-parasites and the so-called **saprophytes** like the bird's nest orchid (*Neottia nidus-avis*) of beech woods
and *Rafflesia* of the eastern tropics. The autotrophic synusia is
subdivided into self-supporting species, climbers and epiphytes.
Richness in lianes (climbers) and epiphytes is another characteristic
of rain forest.

Foresters classify individual trees as *dominant* if they reach above the trees that surround them, *co-dominant* if they share a dominant position, *sub-dominant* if they are below this level but still in the top canopy, and *suppressed* if they are actually below a larger tree's branches. These terms help us to think about the roles of different individual trees in woodland. For example, practically all the fruit is borne by dominant and co-dominant trees, the only ones capable in normal seasons of accumulating an excess of photosynthetic products over their growth and maintenance requirements (see Table 1.1). In some woodlands fast-growing individuals can be picked out by the absence of lichen and algal growth on the stem and branch bark; the reason is simply that rapid accumulation of woody tissues beneath the bark leads to more rapid bark shedding—an epiphytic flora has not time to develop.

The forester's use of the term *dominant* is a rather specialised one, referring to the position of the individual relative to others. Take care not to confuse it with the more general use of the term to describe groups of individuals (usually species populations) which dominate particular ecosystem processes. Thus the botanist picks out as *dominant* those species which make the greatest contribution to vegetation structure because of their size and abundance. He may apply this concept within each stratum above ground. The ecosystem ecologist applies the same idea to all ecosystem processes, noting which species tend to dominate among herbivores, predators and decomposers as well. If by chance one species shows more or less complete domination of a particular process, then a population ecology study of that species alone will give a reasonable estimate of the parameters for the ecosystem process.

It is essential to see the trees and the groundflora as creating in each stratum a wide variety of different *microhabitats* to which other denizens of the wood are restricted in varying degree. Each microhabitat also has a characteristic microclimate which we understand in a commonsense way from our own experience of woodland. Wind-speeds drop; it is cooler on hot days and warmer on cold days; it is sometimes noticeably more humid but seldom drier.

The massive structure of the woodland ecosystem has both advantages and disadvantages for the ecologist investigating eco-system processes. Photosynthetic activity is concentrated in the canopy, decomposition in the litter layer. It is a simple matter to intercept what falls from the canopy and thus estimate a major com-ponent of the turnover of organic matter in the system. Compare this situation with that in grassland where less direct and more time-consuming methods have to be employed. But the massive structure also gives a wide range in the size of pieces of D.O.M. that fall, from delicate bracts to large falling trees. Large branches and whole trees fall sporadically, usually in widely scattered places. An appreciable element in the turnover is thus very difficult to estimate precisely. When such a large piece of D.O.M. falls it creates numerous new microhabitats—another source of areal heterogeneity. The data in Table 1.3 clearly indicate such heterogeneity.

Sample units within the range 0–325 gm^{-2} carry the normal twig fall from standing trees, live and dead. All the others include some material from the crowns of trees felled at the last thinning. Note that the degree of heterogeneity was seemingly reduced when larger sample units were used in 1967. This is a point to remember when interpreting data of this kind and when planning a sampling procedure—for both sets of figures represent essentially the same situation.

So far we have considered only the above ground parts of the ecosystem. Its structure is mirrored to a certain extent underground, the largest trees sending their roots deepest. However the feeding rootlets of all species are often concentrated in the top few centimetres of the soil as soil *pits* or soil *cores* will show. When the soil is so shallow that the massive support roots are also near the surface, windblown trees often reveal the platelike structure of the root system. Equally, with a deep permeable soil and parent material individual roots may penetrate to considerable depths.

Practical work—describing the woodland ecosystem

When the emphasis in a project lies with an animal population, in the groundflora or in the litter and soil, then a general description based solely on visual appraisal may suffice. However the subjective impressions of inexperienced observers are likely to vary considerably. Some quantified descriptive work is then advisable.

1. Does the wood consist entirely of one kind of woodland?
   Can it be overseen from a hill or across the valley? Are there recent aerial photographs? Put in on a map the approximate boundaries of what appear to be different kinds of woodland. Treat each woodland type as a separate entity but discard any that are not relevant to the project.

TABLE 1.3

Random sampling of twig litter in a pine plantation.

The table shows the amount of litter recovered from each quadrat in the chosen sampling areas, the dry weight class to which it belongs and the number of quadrats in each class.

dry weight class g m^{-2}	1966 (sampling area 9.5 m^2 in $\frac{1}{4}$ m^2 quadrats) weight recovered g	frequency	1967 (sampling area 24 m^2 in 1 m^2 quadrats) weight recovered g	frequency
0 – 49	0 5 10 11 11	7	15 24 26 33 47	5
50 – 99	14 16 19 20 22 24	7	66 75 77 85	4
100 – 149	25 26 27 28 31 34 34 36 36 36	11	106 109 133 133 144	5
150 – 199	39 48 49	3	162 183	2
200 – 249	54 57	2	201 235 235	3
250 – 299	66 69	2	275	1
300 – 349		—	325	1
450 – 499	118	1		—
750 – 799	195	1	769	1
850 – 899		—	867 895	2
950 – 999	240	1		—
1150 – 1199	288	1		—
1400 – 1449	359	1		—
2900 – 2949	733	1		—

Otherwise the wood must be inspected on foot and a reasonably homogeneous area selected for study.

2. Describe the location, site topography, drainage and soil in as much detail as appears necessary for the purposes of your investigation.

3. Describe for each stratum the structure and floristics—the general height range of the strata, density and patterns of dispersion of communities or species—correlations between strata characteristics and with variations in site.

The **profile diagram** illustrates the vertical structure characteristics in a strip subjectively selected as representative of the woodland. The length of the strip should be at least twice the general canopy height. Lay out a tape or chain on the ground and note the position of all trees, shrubs and groundflora types. Measure enough heights to ensure that your sketch is properly to scale. If several groups do this independently they are likely to gain some appreciation of the subjective element in this work.

When the project is concerned more directly with the tree stratum then an **enumeration** is usually the first step. In the forester's usage this implies the counting and identification of trees as to species and the recording of one or more parameters such as girth, height, stem taper etc. It is a very time consuming and therefore expensive business. Forestry enumerations are normally restricted to exploitable stems (species of commercial value over a specified minimum girth): they are either exploratory (pilot sampling at $1-2\%$ intensity) or intensive (10–20% intensity). With ecological objectives we must record all species and sizes and it is obvious that in all but the smallest woodland areas we will not be able to make a complete enumeration of the whole area.

Books have been written about sampling. A few of the well tried techniques are offered here. All start by superimposing a grid on the area of the map to be enumerated. It should be orientated so that one line follows the 'lie of the country': this is the **base line** which must also extend to the maximum length of the wood in that direction. The sample transects or transects of plots run at right angles to the base line. A systematic sample runs transects at regular intervals along the base line to the edges of the wood. It has the advantage that all parts of the wood are represented in the sample and it becomes easy to map in the position of paths, windblown patches and areas of differing woodland type. It has the disadvantage that no statistical tests can be run to assess the likely precision of the estimate of the tree numbers and size.

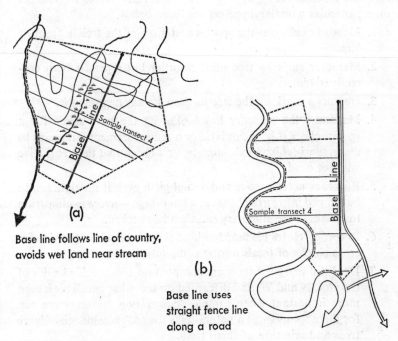

(a)

Base line follows line of country, avoids wet land near stream

(b)

Base line uses straight fence line along a road

Fig. 1.4. Laying out the enumeration grid.

The recommended compromise is to use *stratified random sampling*. Divide the base line into several equal lengths which are multiples of the width of transect or plot you intend using (20 metres is commonly used). According to the intensity of sampling you can afford, choose two or more transects in each section of the base line using random number tables to identify them.

It is rare in work of this sort for the precision of the estimate of the mean to be any better than ± 15–20%. For many ecological purposes greater precision is desirable and you will then find yourself deciding on complete enumeration of what you consider to be a representative area in the wood. A grid within this area will be necessary to ensure that work proceeds systematically. Even now your results are subject to *error*, human error, which can often pass unnoticed and is difficult to check. It is doubly important, therefore, that the work be well organised and meticulously carried out.

As an example of the procedural detail which it is well to stipulate on many kinds of ecological field work, the 'rules' issued to students on a particular woodland project are listed below.

1. Measurer calls out the species and girth of the tree in front of him.

2. Measurer stays by tree until recorder has repeated both correctly to him.

3. He puts a mark on the tree to show he has measured it.

4. Meantime the recorder has looked at the tree, asked for a species check if he is doubtful or a remeasurement if it seems to disagree with his visual estimate, or a check that the tree is alive and not dead.

5. Recorder notes species and actual girth even if the data are to be lumped into girth classes at a later stage—errors in allocation to classes in the summary can then be checked.

6. Recorder 'closes his field book' at short intervals to provide for cross checks of totals during compilation.

7. Doubtful marginal trees are accepted on two specified sides of the plot, N and W, and discarded on the other two. This is even more important when sample enumeration traverses are run for there seems to be a natural bias towards acceptance of large trees and rejection of small trees.

Enumeration data provide quantitative evidence for floristic variation and size structure variation by species over the project area. They are the basis for biomass estimation (see Chapter 6).

4. Describe the structure of the ecosystem below ground.

This is normally beyond the scope of the ecosystem ecologist. The destruction of the habitat is rarely acceptable even if the man-power and the equipment are available. The massive root framework can be revealed by washing away all the mineral material with water jets. The finer roots can be investigated more simply by taking soil cores back to the laboratory, slicing them up at set intervals and washing out the mineral and humic material from each section. In this way the biomass of rootlets per unit volume can be estimated and the levels of the feeding rootlets identified. A disadvantage of working with roots in this way is the difficulty of identifying them as to species. A root investigation therefore has special attractions in a dense wood of one species with little or no groundflora. Ovington (1957) has excavated the root systems of individual trees. One or two studies of massive roots on an area basis have been carried out in North America and much more data should soon become available as the results of the International Biological Programme work are published. If you do have the opportunity to do root excavation, remember that the growth ring analysis method is available for the estimation of biomass increase in large roots (see Chapter 6).

5. Practical work on microhabitats created by the tree stratum.

A little basic information can be obtained without very expensive apparatus. We have seen that consistent correlations can exist between the type of groundflora and the amount of light penetrating to the woodland floor as measured with a light meter of the kind used in photography. An overcast day is much to be preferred for this kind of work because there will be no sunflecks to raise, manyfold, the variation between readings at a given spot and it does not matter so much whether the meter is held exactly in a horizontal plane.

In a secluded part of the wood it may be practicable to put out maximum and minimum thermometers to compare the temperature range in gaps of various sizes and under full canopy. Humidity may be a critical parameter for some organisms. Cloudsley–Thomson (1967) describes some of the more sophisticated instruments that are available for measuring humidity, temperature and light regimes in microhabitats.

2 How do woodlands change with time?

[handwritten annotations: "time of year could effect organisms in the soil"]

It is often suggested that areas of tropical rain forest have existed unchanged for millions of years. This is almost certainly an exaggeration, but it does make the point that we think of change in very general terms. We do not imagine that the detailed structure in one small area remains the same. Trees must grow old and die or be blown over. Seedlings must grow up to replace them. Locally species composition will change while, over a large enough area, change may appear negligible. Fluctuations in time about a relatively unchanging norm constitute the concept of *dynamic equilibrium*. These fluctuations are conveniently categorised as diurnal, seasonal or longer term.

The daily cycle of change in woodland stems from changes in the physical environment. Energy is reradiated at night and the wood becomes cooler. In the early morning moisture tends to condense out from the atmosphere. By noon the highest temperatures and lowest humidities are recorded above the canopy. Shadows move across gaps as the sun falls below its zenith. Each population is regulated in its activities by this cycle. The trees begin photosynthesis very early but by noon on warm days stomata will be closed, slowing down both photosynthesis and water loss. Cold blooded animals (*poikilotherms*) will not stir till their immediate environment has become warm, often late in the morning. Warm blooded animals (*homoiotherms*) such as deer may graze at dawn and dusk but lie up at mid-day.

Seasonal changes are more striking especially in deciduous woodland. In autumn leaves fall and more of the incident light reaches the ground. But the sun is weaker and reradiation of heat absorbed during a shorter day is stronger. Woodland becomes a much harsher place for its denizens. The above ground parts of perennial herbs die back. Many animals hibernate or overwinter in their inactive phases as eggs, cysts or pupae. Annual plants survive as seeds or spores. In such deciduous woods we associate spring with carpets of primroses (*Primula vulgaris*) or bluebells (*Endymion non-scriptus*). These are plants adapted to exploit the early period in spring before the tree leaves are fully unfolded.

Ecosystem processes are closely linked to the seasonal cycle. Species of the higher trophic levels must, like the early spring flowers, be adapted to exploit the energy sources available to them. Generally they cannot become very active or very abundant until their food source has become plentiful. Thus the main growth period of autotroph populations is spring and early summer—of herbivores, late spring and summer—of predators, summer—and of decomposers, late summer and autumn. The nutrient cycle is annual, its component processes, *uptake*, *turnover* and *decomposition* reaching peaks in successive seasons. Compare this situation with that in the humid tropics where there are no seasons. Ecosystem processes are not phased through the year: the nutrient cycle is completed in four to six months.

There are often noticeable variations in seasonal patterns from year to year, particularly in Britain where the vagaries of the macroclimate are notorious. Wet summers bring forward the period of peak fungal activity from October to August. September frosts accelerate leaf-fall and late spring frosts can destroy so much young foliage that the prospect of a tree achieving much net *primary production* in that year is poor. Thus mast years of beech (*Fagus sylvatica*) tend to follow years with particularly favourable growing seasons when primary production more than meets the maintenance requirements. Generally the productivity of an ecosystem will fluctuate from year to year above and below what is considered to be a characteristic level. Only when this characteristic level changes—and this can only be assessed over a period of years—is it suggested that any real change is occurring in the wood.

The populations that make up the biota of an ecosystem tend to be of two kinds—those with an all-age structure and those with little or no age range. Human populations have an all-age structure and this has perhaps influenced our concept of what is the norm. Certainly it applies to the mammals and birds we find in woodland. But smaller animals and micro-organisms often have life-spans of a year or less. Activity is synchronous and the populations are more or less even-aged. Annual flowering plants fall in this category and so, too, do most of the trees of our woodland. Whether this is a 'natural' or an 'unnatural' phenomenon will be discussed at some length in the next chapter.

For the moment let us pursue the implications of the fact that the tree species populations which dominate our woodlands are more or less even-aged over appreciable areas which we will designate as **stands**. The stand ages as the individuals age together, their life cycle characteristics determining some of the features of a **stand-cycle**. Watt (1947) first used this term in connection with heath vegetation, describing four characteristic phases. Below they are described as they apply to woodland:

1 *Regenerative phase*—young seedling trees, scattered but sometimes with locally dense patches. The vegetation is open at this stage, allowing the establishment of many adventitious species among the tree seedlings.

2 *Building phase*—the young sapling trees have grown laterally into contact with each other. The canopy is dense. The shade-intolerant species associated with the regenerative phase are gradually eliminated. This is the period of maximum biomass accumulation (hence the name). It is also the period of intense intraspecific competition. Many individuals become suppressed and die at an early age while their more successful neighbours expand their crowns above.

3 *Mature phase*—competition is no longer so intense. The rate of biomass accumulation in the stand has begun to slow down. The crowns of the trees are healthy and bear their heaviest crops of seed at this time. However the canopy becomes less dense than in the building phase and a shade tolerant groundflora becomes established.

4 *Degenerative phase*—this is the period of senescence during which biomass accumulation is initially small and eventually negative. The aged trees may become stag-headed as branches die. Large branches decay and fall, whole trees die and fall. D.O.M. accumulates in the ecosystem as the biomass declines. Light reaches the ground in increasing amounts.

This last phase is seldom seen in Britain. Woods are harvested before heart-rots and other decay organisms reduce the quality and quantity of timber. Indeed the forester thinks of woods as degenerating as soon as current stem production falls below the average production to date (Fig. 2.1). This is almost certainly still within the mature phase of Watt.

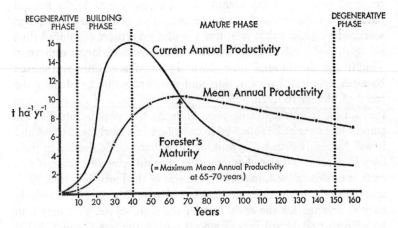

Fig. 2.1. Productivity changes with age in a Scots pine stand.

Management practices have other effects on the stand-cycle as described above. For example, the forester attempts to eliminate as much as possible of the self-thinning that occurs during the building phase: firstly by planting at regular spacing fairly far apart; secondly by thinning the stand at regular intervals, limiting the amount of suppression that occurs and harvesting stems before they die.

What happens in the degenerative phase is thus largely a matter of speculation. Visible evidence of senescence may not appear in a pine stand till it is 150–200 years old. Individual trees will obviously have very different life-spans and it would seem that patches of the regenerative phase should appear within an old stand. Gradually the stand should develop an all-aged structure. Why this has seldom happened in Britain or, indeed, in most parts of the world is discussed in the next two chapters.

For the moment we will merely note that a woodland stand passes through four major phases, sufficiently different for each to be occupied by different sets of associated species of plants and animals.

Practical work—seasonal and short-term change

A great deal of routine observation of change in woodland is required before you can identify those changes outside the normal pattern of variation and therefore indicative of extrinsic factors at work. The odds are that your tree population is more even-aged than all-aged, that it will be growing older, the individuals larger each year. Thus in addition to seasonal cycles and yearly fluctuations you should be able, by careful observation and measurements, to identify the trend of change in the groundflora and other parts of the ecosystem that follow from the ageing process in the tree population. But the stand-cycle cannot be followed in one place in a working lifetime and it will be well worth while looking for other woods similar to your own in species and site characteristics, but differing in age or, if this cannot be determined, in the average size of the individual tree. With so much replanting since the last world war, this may be relatively easy to arrange for the early part of the stand-cycle. You may also be able to find broad leaved woods within the age range 120–220 years; for a great many of the woods in and around our cities and towns and on the surviving big estates were planted between 1750 and 1850.

'Permanent' plots or transects will be needed—a combination of the two if you attempt to monitor all the strata. They should be demarcated inconspicuously with small creosoted pickets barely projecting above ground. These pickets will become covered with litter and it is essential that you tie in each plot or series of plots to a conspicuous and 'permanent' feature in the wood.

Changes in the tree stratum may be assessed by repeated enumeration (Chapter 1) and the continuing analysis of litter fall (Chapter 6). There are several methods of assessing groundflora changes.

Sketching permanent quadrats at intervals Figure 2.2 illustrates a situation in which this method works quite well. The project from which these examples were taken actually concerned the development of natural regeneration of tree species in the Old Wood at Cawdor, Nairn. It was abandoned after three years when it became apparent that deer browsing was preventing any of the tree seedlings from growing much above 25 cm high.

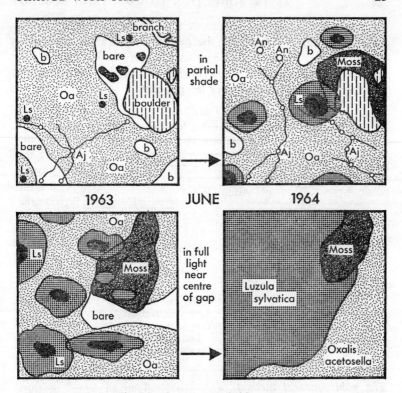

Fig. 2.2. Changes recorded in a year.
Permanent quadrats in Cawdor Old Wood near Nairn—area 1 m².
Tracing the changes in groundflora following the felling of a large beech
tree which left bare areas (covered with litter but no plants).

Aj	*Ajuga reptans*	An	*Anemone nemoralis*	Mosses included
Ls	*Luzula sylvatica*	Oa	*Oxalis acetosella*	*Mnium hornum,*
b	bare ground			*Thuidium tamariscinum*
				and *Rhytidiadelphus*
				loreus

Domin Scale assessment Domin appreciated that it was difficult to
assess cover per cent accurately by eye. He therefore prepared his
widely used assessment technique in which cover per cent is estimated in eight classes described in Table 2.1 overleaf.

The smallest class is subdivided according to frequency. A major
disadvantage of the **Domin Scale** is its hybrid nature and the irregular
range of its cover classes: for you cannot sum Domin Scale numbers
and arrive at a meaningful average. Bannister studied the relation-

ship between cover and frequency and affected a transformation of the Domin Number into the additive cover/abundance index shown above. When beginning this kind of work a metre stick marked off in decimetres is a useful aid in visualizing the area within a quadrat that represents five or ten per cent.

TABLE 2.1

The Domin Scale and Bannister's Cover/Abundance Index.
(from Bannister, 1966).

Domin Scale No.	Cover		Bannister Index No.
10	95–100%		83
9	75–95%		74
8	50–75%		59
7	33–50%		46
6	25–33%		39
5	10–25%		31
4	5–10%		26
3	<5%	Numerous individuals	9
2	<5%	Several individuals	4
1	<5%	One or two individuals	2
+	0	Present in surround	1

The point quadrat method Visual assessment of cover becomes increasingly difficult in grassy vegetation or tall herb vegetation. Then the so-called **point quadrat** method is indicated. It works on the principle that if a very large number of pins descend on vegetation from above, the number of hits on different species will be proportional to their cover. If all hits are recorded until the pin reaches the ground (multiple contact point quadrat) the results show quite good correlation with above ground biomass in some vegetation types. In theory the pin should be infinitely thin for a pin can only detect gaps in cover that are bigger than its diameter. In practice this bias must be accepted for a slender pin is readily bent and may be dangerous.

The method will work in any vegetation of moderate height and it can be used to check visual estimates of cover. With 100 points in a metre square the cover of the major dominants may be estimated more accurately than by eye but the species with low cover tend to be over or underestimated and some species are not recorded at all. A

minimum of 500 points is probably required to achieve a reasonable
cover estimate for most of the species present.

TABLE 2.2

Summary of point quadrat assessments in the groundflora of an alder wood
felled in the winter of 1964.

June	1964	1965	1966	1968	1969
No. of points	500	500	615	600	500
Trees and shrubs	5·6*	6·2	8·8	17·9*	26·2*
Dicot herbs	77·6	49·7	54·4	33·1	31·2
Grasses	6·6	38·1	31·2	47·0	36·0
Other monocots	0·4	0·7	1·0	1·6	4·8
Ferns	4·8	2·8	3·3	3·8	1·4
Bare	5·0	2·5	1·3	0	0

* The alder overwood in 1964 was more or less complete: the 1968 and 1969
figures include increasing cover from alder coppice shoots.

The above data were obtained using a very crude piece of apparatus
—a metre stick with five fixed pins lowered on to the vegetation by
hand at one metre intervals along three transects. The transects were
not in the same place each year which accounts for some of the
variability shown. The figure for grasses in 1965 is almost certainly an
overestimate arising because the pins were thick and the panicles of
Poa trivialis were fully expanded.

A point quadrat frame for low vegetation can be quite simply constructed with two spaced cross pieces drilled with holes at 5, 10 or
20 cm intervals: the holes are made just large enough for steel
knitting needles to slip through without too much resistance. If
longer pins are needed 3 mm steel wire rods can be cut. An apparatus
with spiked 1·5 m uprights and adjustable level cross pieces is not very
expensive to construct if you have the facilities: several sets of pins
could be cut for use in vegetation of different heights.

3 Succession and climax

New surfaces devoid of life are created around us every day—a deep cutting for a motorway—a new sandbank in the river after a heavy storm—a new wall or mine tip. Such places are usually inhospitable and only a few hardy pioneer species of plant can colonise them; perhaps minute algae or lichens, sometimes larger plants. Organic matter from the dead parts of these plants mixes with mineral matter of the parent material binding it together and providing a food source for pioneer decomposer organisms. This microflora, and the micro-fauna that follow, convert the sterile substrate into soil. The soil volume increases by root penetration below and humus incorporation from above and within. The soil moisture regime is improved. Sands become more retentive of moisture. Silts and clays develop a structure which allows freer drainage. Plant nutrients etched from the parent material are less readily leached away or become more accessible: *nutrient capital* accumulates. Thus pioneer plants upgrade the ecosystem making it suitable for more exacting, less hardy, species. The pioneers cannot compete with species better equipped to exploit the improved conditions and so the pioneers disappear. This replacement process repeats itself many times during ecosystem development. Botanically the process is called **succession** and the whole development a **sere** in which the major *seral stages* are recognised from the most prominent plant species present at the time. The early pioneer species tend to be short lived. If they are flowering plants they will be annuals. As vegetation develops perennials of increasing longevity take over dominance.

Thus the early period within a sere when conditions are harsh shows slow change. Equally the late period shows slow change because the dominants are of great longevity. In the moist tropics all the dominants of a sere may be trees e.g. mangrove succession on newly exposed mudbanks out to sea. In our part of the world woodland succession is associated with the later seral stages only.

26

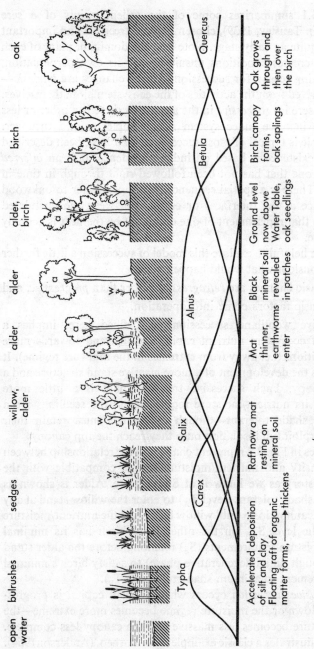

Fig. 3.1. Hydrosere at Sweat Mere in Shropshire (from: Tansley, 1939).

Figure 3.1 summarizes some of the salient points of a sere (Clapham in Tansley, 1939), chosen because trees play an important part from quite an early stage. Note how the dominant plant of each seral stage creates conditions unsuitable for its own perpetuation, the underlying reason for succession. All the changes are *intrinsic*— they are derived from the activities of the ecosystem biota themselves. The whole sere is *progressive* in the sense that there is a more or less continuous increase in ecosystem size, complexity and turnover. This example is typical of most successions that have been described in that its existence is based on indirect evidence. It is an *inferred succession*, one that has not been followed right through in time in one place. There is a spatial sequence from open water to oakwood and evidence at the interfaces or **ecotones** between the postulated seral stages that replacement of one dominant by the next is actually taking place.

It may be helpful to explore this model of succession a little further and then consider what would happen when

(*a*) intrinsic changes are *retrogressive* rather than *progressive*, and

(*b*) extrinsic factors come into operation.

Progressive woodland succession, as defined here, implies a build-up of ecosystem nutrient capital and a trend towards more mesic conditions (i.e. away from extremes in the moisture regime). It also implies the development of a more massive stand structure and a denser canopy. Each successive tree dominant is a little more exacting in its nutrient/moisture requirements: the seedlings are a little more shade tolerant—though individuals must retain their ability to exploit full insolation once they reach the top canopy.

The curves in Fig. 3.2 represent changes in the relationship between canopy density and nutrient/moisture regime compatible with the succession story as we know it at Sweat Mere. Alder is shown as sufficiently shade tolerant (level T_A) to enter the willow stand at any stage but it cannot do so until willow has raised the nutrient/moisture status to the level S_A. On the other hand birch has its minimal nutrient/moisture requirements (S_B) met long before the alder stand is open enough (in its degenerative phase) to satisfy birch's minimal light requirement (maximum shade tolerance T_B).

Retrogressive succession occurs when nutrient capital is progressively run down or the moisture regime becomes more extreme—the stand structure becomes less massive and the canopy less complete. Figure 3.3 illustrates a classic example from Borneo, (Anderson 1964).

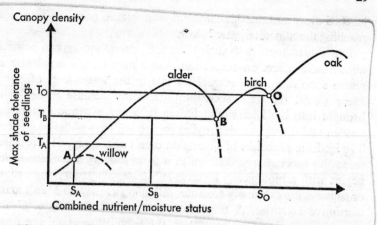

Fig. 3.2. The relationship between canopy density and nutrient/moisture status during succession.

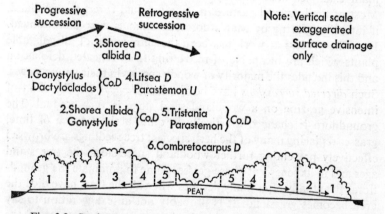

Fig. 3.3. Section across deltaic island at Baram, Sarawak (from: Anderson, 1964).

This is another spatial succession, progressive up to seral stage 2 and then retrogressive to the open vegetation dominated by shrubby individuals of *Combretocarpus* in seral stage 6. Pollen analysis of a core taken in the centre confirms that all the earlier seral stages have existed in sequence at this point. There is an exact parallel in raised bog formation in these islands. The intrinsic reason for retrogression is the accumulation of peaty remains in a heavy rainfall climate: as the 'ground' level rises in the centre nutrients are progressively leached away and the thick layer of waterlogged peat prevents roots reaching the mineral matter below.

Extrinsic factors are so varied that few generalisations can be made about their effects on succession. Some may merely accelerate or retard succession. For example, a rise in the water level of Sweat Mere would set back succession while an increase in sediment brought into the lake would hasten it. Some extrinsic factors are sudden and disruptive, destroying or altering part of the ecosystem. The intrinsic processes of succession then begin again. This is called **secondary succession** to distinguish it from **primary succession** which begins with a biologically inactive substrate. Secondary successions or **subseres** are extremely common following as they do Man's many disruptive activities. A third category of extrinsic factors diverts normal successions by preventing the establishment of the usual dominants. Such factors must be operating continuously. Again the main examples arise directly or indirectly from Man's activities. Most of the low level pasture in Britain would revert to woodland, if intensive grazing by Man's domestic animals were discontinued. Here the process at work has been the elimination of all palatable plants which are incapable of recovering from repeated defoliation and this includes the majority of woody species in their young stages. Such *diverted successions* have been called *plagioseres*. The effect of intensive grazing on a woodland seral stage is very gradual. The groundflora is changed drastically in quite a short space of time, grasses replacing many of the herb species: tree seedlings are browsed effectively preventing further woodland succession: the tree stand ages and in its senescent phase grasses come to dominate the ground-flora. Eventually after many decades when all the trees have died the area becomes grassland. It is probably not an exaggeration to say that sheep are responsible for the low 'tree line' in most of the mountainous areas of Britain.

Figure 3.2 shows succession as an open-ended system. Once the dominant is a long-lived species such as oak, successional change is very difficult to observe. The dimensions of change are less than those of the seasonal and yearly changes that occur in any ecosystem in response to fluctuations in the abiotic and biotic environment. How can we tell whether the sere has reached its **climax**? This is a question that has intrigued ecologists for nearly a century. When woodland dominants may live for 300 years or more we can hardly wait to see what will happen. Of course, if we know that a woodland area has

remained substantially unchanged for many centuries, that is fair
evidence that the climax has been reached. In the *hydrosere* at
Sweat Mere, oak is the most exacting tree species we know in this
kind of climate and terrain. We cannot envisage any other species
replacing it and again this is fair evidence that the climax has been
reached. But is it fair to assume that the most exacting species the
climate allows will always form the climax? Within the lowland oak
woodland belt of northern Europe, there are, locally, extensive pine
forests. They have existed for many centuries. Their soils are derived
from course sands and gravels, parent materials which are readily
leached of the small amounts of plant nutrients that can be derived
from them. It seems reasonable to conclude that these are indeed
climax forests in which the nutrient/moisture status cannot be built
up by intrinsic processes to the minimum level required by oak. There
are therefore lowland *edaphic climaxes* of pine in the zone of *climatic
climax* of oak. Beech '*hangers*' on the steep slopes of England's
chalk and limestone hills provide another example of an edaphic
climax.

It has seemed that there should be an even more reliable criterion
of climax in the age structure of the stand itself. Seral stages are
identified by the absence of young individuals of the reigning dominant; for it creates conditions unsuitable for its own regeneration.

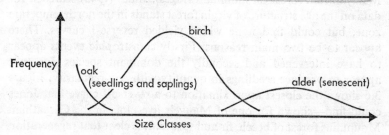

Fig. 3.4. A seral situation in which an analysis of size classes shows that
birch has replaced alder and is going to be replaced by oak (cf. Fig. 3.1.).

A climax stand, on the other hand, must persist with the same
dominant—there must always be young individuals to replace the
old ones as they die. Analysed by age classes, frequency should fall
sharply at first and then more slowly among the older age classes,
giving a reverse-J curve. Figure 3.5 overleaf shows the characteristic
reverse-J curves of tropical rain forests.

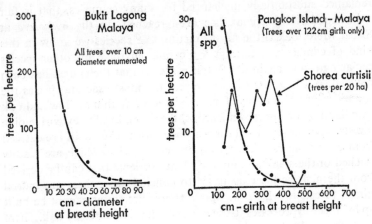

Fig. 3.5. The reverse-J curve for all species in tropical rain forest (Bukit Lagong data from: Wyatt-Smith, 1949).

Unfortunately rain forest trees are not easily aged (see Chapter 2) and there is extreme diversity of species. These curves are for size-classes, not age classes, and for all species combined. Although some species follow the general trend, there are others which do not (see the parameter for *Shorea curtisii*, one of the giant emergent trees of Malayan rain forest on mountain ridges). Jones (1945) searched for data on the age structure of virgin forest stands in the north temperate zone, but could find none which yielded reverse-J curves. There appear to be two main reasons. Firstly catastrophic events appear to have intervened and secondly the dominant species does not appear to produce seedlings in quantity with any regularity. Figure 3.6 shows the closest approximation to a reverse-J curve that Jones could find—shown for one of Mauve's five plots in the Carpathian Mountains forest of beech, fir and spruce. It is clear that regeneration of each of these species comes in waves, corresponding to periods when seed production was prolific and conditions were at the same time particularly suitable for germination and development. If this is climax forest, as it seems to be, then there must be alternation of dominants with time on any one area.

Virgin forest areas have been virtually eliminated from Britain. There are some little disturbed remnants of ancient pine forest in Scotland. None of these are all-aged. There is one predominant age class in many of them and in the others two widely separated age classes. They appear to have developed after fire had completely or partially destroyed the previous pine stand. Now, in Britain, the human fire hazard has so dominated the scene, that it was a long while before we appreciated that an abiotic fire hazard existed. Today in Finland six or seven fires a year in every 10 000 hectares of forest are attributed to dry lightning storms. Destruction of forest by fire is widespread in the boreal forests and also in areas with low summer rainfall further south. Where fire is a recurrent natural hazard the term *fire-climax* has been used for forest types which appear to be maintained as such in this way. The climax passes through repeated stand-cycles, for fire recreates conditions suitable for regeneration of the dominant. In terms of our general successional theory (see Fig. 3.2) note that the complete destruction of all above ground structure will allow earlier seral dominants to establish themselves. And, indeed, they are commonly present in the re-generative phase of pine.

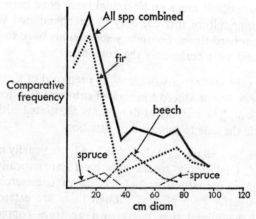

Fig. 3.6. Mauve's Plot A. Northern slopes of the Carpathian Mts. (from: Jones, 1945).

Elsewhere in the north temperate zone aberrant hurricanes and epidemic disease periodically lay waste areas of forest. It would appear that disruptive agencies have recurred in the same place at intervals less than the times needed for woodland seres to reach their climax. In central Europe where this statement may not apply, waves of successful regeneration seem to appear at long intervals: again the result is forest in which a species is represented at any one place by one predominant age class. A universal criterion for woodland climax remains elusive and it must be questioned whether the concept is a valuable one today, particularly when the impact of Man on woodland has still to be considered.

Man has affected successions and climax in many ways. We have already noted some of the effects of intensive grazing by domesticated animals. In the next two chapters we shall be looking at the overall effects of Man's activities. They all affect successions and climax:

1. Clearance, usually for agriculture, has broken up most of the forest into small relatively isolated areas of woodland. Tree seed sources are dispersed and seed of dominants may not become available at the appropriate time in a succession. For example, the birch seral stage at Sweat Mere might have been dominated by the more massive black poplar had there been enough seed-bearers of the latter in the neighbourhood.

2. Many woodland areas on 'marginal land' have been cleared in times when agriculture thrived, only to be abandoned when agriculture fell on hard times. Secondary successions back to scrub and then woodland were commonly the outcome.

3. Partial exploitation of woodland, the removal of single trees or groups of trees, has produced a mosaic of sub-seres at different stages of development. The mixture of dominants associated with different seral stages is the clue to this kind of situation.

4. Managed woodlands, on the other hand, are usually notable for their 'purity'. Undesirable species have been systematically removed and there is an unnatural preponderance of commercial species. They have usually been planted though there are extensive areas, still, of oak woodland that has grown up from coppice. Where indigenous species have been planted ecological interpretation may become very difficult for they may not have been planted in areas to which they are known to belong. Thus beech has been planted extensively north of the Midlands and pedunculate oak in the north and west of Britain on sites where sessile oak would normally occur.

With the exception of partial exploitation, all these activities lead to woodland stands of more or less even age and woodland groundfloras must vary with the stage in the stand-cycle as well as with the stage in succession. Remember, too, that the groundflora may be subject to very varied intensity of grazing or human trampling and you will realise that the situation is usually very complex indeed at ground level. Nevertheless, there is a chance that characteristic groundfloras of Britain's climax and seral woodland types have persisted locally in areas that have never been under agriculture. In Britain, then, human interference has been so extensive that it becomes imperative to learn all we can about the history of an area before attempting to define the status of its woodlands.

Practical work—succession

Succession in woodland, as we have seen, is a slow process and correspondingly difficult to detect in one place over a short period. The best opportunities for demonstrating the kind of situation described in Figure 3.4 are in mature stands of pioneer species such as hawthorn (*Crataegus monogyna*), willow (*Salix capraea* and other shrubby willows), birch, alder and sometimes Scots pine (pine if you remember may be an edaphic climax species and in a terminal stand cycle).

A complete enumeration of all stages is required, including seedlings and saplings. The aim is to plot the frequency of quite broad age classes, perhaps 20 year classes for species of greater longevity and 5–10 year classes for shrubs. New seedlings may be put in a special category for they tend to fluctuate markedly from year to year. In practice you will probably record size classes and then try to establish a general curve for size upon age by counting the rings in a small number of stems covering the size range present. Remember to add in the time a plant takes to reach the height at which the ring count material was taken. The area you need enumerate to obtain a representative sample will vary with the size of the larger trees and the homogeneity of the stand. It will be convenient to decide on an arbitrary cut-off point for enumeration of larger individuals: sub-samples (systematic strips or transects of plots within the enumeration area) will suffice for the smaller individuals below the cut-off. The tree parameter measured will usually be height for the smaller individuals and girth breast height for individuals above the cut-off dimension.

When an apparent succession exists in space, as at Sweat Mere, rather more complete data can be obtained. If the distance is short, try a complete transect closing the record at regular intervals less than the width of the least extensive seral stage. Otherwise a transect of spaced plots is indicated: these should be subjectively sited to obtain representation of each full seral stage and the transitions between. It will be routine to check your inferred succession in any way you can. Had you worked at Sweat Mere, for example, you would have dug soil pits in the birch and oak dominated zones to see whether the soil structure there was consistent with an origin in lacustrine silts.

Domestic herbivores often have access to woodland for browsing and shelter. The processes of succession in your woodland may have been diverted (plagiosere). If so, the groundflora may bear little resemblance to that of otherwise similar woods that have been protected. Indeed you should, as a matter of course, look out for evidence of browsing and grazing before attempting to interpret data in which the young age classes are poorly represented. Confirmation of suspicions that herbivores are affecting succession can often be obtained quite strikingly and in a surprisingly short period by fencing a small area.

Before attempting to allocate trees of various size to even broad age classes it might be salutary to study the range of sizes present in a plantation of known planting date.

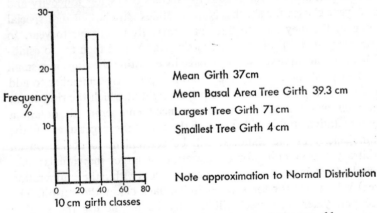

Mean Girth 37cm
Mean Basal Area Tree Girth 39.3 cm
Largest Tree Girth 71cm
Smallest Tree Girth 4 cm

Note approximation to Normal Distribution

Fig. 3.7. Spread of size classes in a pine stand 39 years old.

Note in the field the difference in appearance between trees that are small in girth because they have failed to get up to the top canopy (suppressed individuals) and trees of similar girth in the open with heavy crowns and a vigorous look about them. Obviously girth is a poor indicator of age, but there is a limit to the amount of felling, if any, that can be done to determine age. You may sometimes be able to use a Pressler Borer which will give you an indication of the speed of growth during the time taken to lay down the outer 3–4 cm of wood on the stem. In pine stands up to a moderate age, a count of the whorls will tell you how old a tree is. With older trees this is not possible, partly because of irregular growth at the top and partly because the remains of all the lower whorls will have been grown over leaving no signs at the surface.

4 The history of British woodland

A million years ago a comparatively sudden lowering of temperature was the harbinger of the Ice Age. Fluctuations in temperature became more extreme and at intervals of the order of 100 000 years the polar ice-cap grew enormously in size. The water locked in the ice led to a depression of the sea-level all over the world as the ice pressed southward overland and brought an Ice Age to northern Europe. During the warmer periods the ice receded northwards in what are called the *interglacial periods*—shorter period fluctuations of warmer climate are *interstadials*. The most extensive incursions of ice into Britain reached a line from the Thames to the Severn estuary. The most recent major glaciation began about 70 000 years ago reaching, at its height, the northwest Midlands and what are now coastal areas around the Wash.

We do not know what cosmic phenomena cause the Ice Ages. We cannot know when, or even if, they will cease. It is a sobering thought that in another 10 000 years glaciers may again be descending from the Scottish Highlands. Certainly the weight of evidence suggests that we are living in a major interglacial—it has been called the Flandrian interglacial.

In all the interglacials to date there has been a similar pattern of climatic and geomorphological change, see Fig. 4.1.

We would know little about the past vegetation in Britain of these

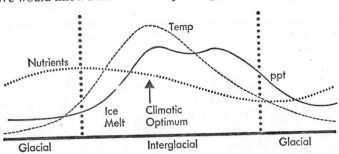

Fig. 4.1. The general climatic and geomorphological pattern of an interglacial.

times were it not for the fact that pollen grains have
(*exine*) extremely resistant to decay. When pollen fell i
peat bogs, lakes and estuarine muds, it has been preserved
sable form. Careful analysis, layer by layer, gives some i
plants that were abundant in the neighbourhood when ea
was laid down. *Palynology* is the name given to the science t ...as
developed from such studies. The many problems involved in inter-
pretation of the pollen record cannot be discussed at length here. We
may note that species vary in their pollen production, in the durability
of their pollen grains: some species will have grown close to the
wet area in which the pollen was preserved, others much further
away: there are often discontinuities in deposition which make dating
difficult. Nevertheless patterns of change in the amounts of pollen
of various species with strata (and therefore with time) in any one
deposit tend to be reinforced by similar patterns from other deposits
in the same region.

Figure 4.2 overleaf allows a comparison of reconstructions of
vegetational changes in all three interglacials. In all three, birch
(*Betula* spp.) and pine (*Pinus silvestris*) are most abundant towards
the beginning and end. Oak (*Quercus* spp.) comes in as the tempera-
ture approaches the maximum. Alder (*Alnus glutinosa*) is only
abundant once rainfall has reached its higher levels. Spruce (*Picea
abies*) comes in only in the later stages.

With each ice incursion vegetation was destroyed on all except the
occasional hill or mountain top which escaped the ice. The south of
England had the cold, dry (*periglacial*) climate characteristic of areas
at the fringe of the ice-cap. Trees could not survive. Thus during each
interglacial, trees had to recolonise Britain from areas far to the south
and east. Each time one or two tree species failed to get back to
Britain and our indigenous tree flora has become progressively
impoverished. Silver fir (*Abies alba*) was present in the Hoxnian
interglacial; Norway maple (*Acer platanoides*) and spruce were
present in the last completed interglacial (the Ipswichian).

In the same way the diversity of shrubs, herbs and animals in our
woodland must have been progressively reduced, though there is not
much direct evidence. The existence of a major east-west mountain
chain in Europe has exacerbated the situation; for species could not
migrate easily to and from refuges south of the Alps and Pyrenees.
Compare the far richer tree flora of temperate North America where
the main mountain ranges run north-south.

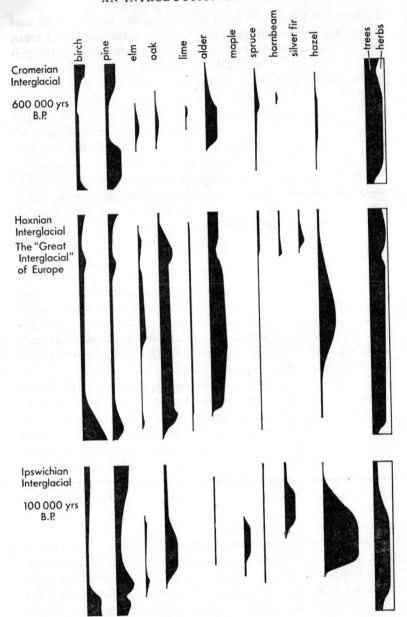

Fig. 4.2. Schematic pollen diagrams for three interglacials (from: Pennington 1969).

There are several points of ecological importance to us today which arise from this period of our woodland history :

1. Several exotic tree species have been shown to belong in Britain in the sense that they have been indigenous under conditions like the present in previous interglacials. Where they have been reintroduced by Man, as all of them have, we may expect that in due course they will become fully naturalized again.

2. Because our woodland flora is so impoverished we must expect that British representatives of species will have broader **niches** (see Chapter 5) than their counterparts in Europe. Some of the implications of this situation will be considered in Appendix II.

3. The pattern of vegetation change in interglacials appears to be predictable. Let us look, then, at the changing vegetation of Britain since the last Ice Age and see whether it conforms to the expected pattern, Table 4.1 overleaf.

The tree birches (*B. pubescens* and *B. pendula*) were the first to form areas of woodland—in the Allerød period (which we could classify as a minor interstadial) and then again in the Pre-Boreal after an intervening period of herbaceous vegetation. Undoubtedly there was climatic limitation of succession, though the situation is complicated by the possibility that later seral dominants did not arrive as soon as conditions were suitable for their establishment. We see that early seral species, birches then pine, were dominant for nearly 3 000 years up to the end of the Boreal period. Locally, and probably on the better soils only, elm (presumably *Ulmus glabra*) and oak (presumably both *Q. petraea* and *Q. robur*) were becoming well established. But conditions generally were not sufficiently favourable for them to become the prevailing climax dominants: in the very north of Britain they never did.

The Boreal was a period of increasing warmth, increasing rate of ice melt and corresponding rise in sea-level. About 7 500 years ago the sea finally cut through the chalk hills near Dover and made Britain an island. In so doing it made the northward emigration of species from the Continent even more difficult. The woodland species to arrive in Britain since this time have nearly all been deliberately or unwittingly introduced by Man. The spread of the sea into the low-lying ground that is now the North Sea and the Irish Sea accentuated the trend towards a more oceanic climate. The

Atlantic period, which followed the Boreal, saw the continuation of the trend to greater warmth but it was now accompanied by heavy rainfall. These were the conditions under which *mixed oak forest* (M.O.F.) became the prevailing climax vegetation through most of lowland Britain. At the end of the Atlantic the *Climatic Optimum* was reached and thermophilic tree species such as the limes (*Tilia platyphyllos* and *T. cordata*) made an appreciable contribution to the pollen 'rain'. Alder became abundant on the much more extensive areas of marshy ground.

Up to this point, some 5000 years ago, the pattern appears to follow faithfully that of the previous interglacials. But it was also just about the time that the Neolithic peoples with their flint axes arrived in this country. From now on the expected interglacial pattern is increasingly altered by forest clearances and cultivation. Bronze Age and then Iron Age peoples introduced more efficient axes, the hoe and the plough. At first, shifting cultivation was practised. Woods were felled and burnt leaving charcoal layers in the soil. The pollen record nearby faithfully relates the story, with cereal and weed pollen followed in higher layers by tree pollens as secondary succession took place when the cultivation was abandoned. Grassland and heath were becoming more extensive as pastoral farmers felled and burned woodland to increase their grazing areas. Dimbleby (1952) has produced convincing evidence that much of the heath and moorland in Yorkshire was woodland in the early Sub-Boreal. A marked decline in elm pollen during the Sub-Boreal is now thought to be due to Neolithic Man's practice of lopping elm branches for cattle fodder. Towards the end of the Roman occupation (about 300 A.D.) it is estimated that there were some two million (out of fifty million) acres in cultivation with a rather larger area in pastoral use. The main massifs of forest were still intact.

The Saxons introduced farming techniques capable of exploiting the heavy clays on which much of the M.O.F. stood and by the time of the Norman Conquest there were some ten million acres in cultivation. Deer and wild pig were still abundant as were wolves in the north. Oak and beech woodlands were regularly used for *pannage* of swine which suggests that yearly crops of acorns and beech masts were the rule; that the climate was then more favourable for these species than it is today.

This is a reminder that climate is continually fluctuating over short, medium and long-term periods. Lamb (1972) has done some

TABLE 4.1

Summary of events of ecological importance in Britain since the last Ice Age (from: Godwin, 1956)

Time Scale B.P.	Climatic periods	Major datum horizons in bogs	Pollen zones	Principal tree spp.	Forest cover	Notes	Time scale
Present	**Sub-Atlantic** cool and wet		VIII	oak birch alder (beech)	Cleared by Man	Little Ice Age Golden Age of British Agriculture Anglo-Saxons Romans Iron Age Man (ploughs)	1670–1700 AD 13th Century 400 AD 50 BC 500 BC
	Sub-Boreal declining warmth increasingly dry	Regeneration of bogs Bog growth ceases	VIIb	Mixed Oak Forest (alder)		Bronze Age Man (hoes)	1 800 BC
2 500	*Temp. Maximum Climatic Optimum*			lime max.	FOREST		
5 000	**Atlantic** increasing warmth and wetness		VIIa	elm max. M.O.F. alder		Neolithic Man (flint axes) Mesolithic Man (hunter)	3 000 BC
	Boreal increasing warmth and dryness	Regeneration of bogs Bog growth ceases	V IV	hazel-pine		British Channel formed Ireland becomes an Island	5 500 BC ?
7 500							
9 600	**Pre-Boreal**		III II I	birch birch		North Sea floor still dry	7 600 BC
10 200 Late glacial	**Upper Dryas** **Interstadial** **Lower Dryas**			birch (juniper)	open vegetation	Sub-arctic conditions	8 200 BC

fascinating detective work into past climate. It is clear that since Roman times temperature was in a fluctuating uptrend culminating in the 13th century—the so-called Golden Age of Agriculture when vineyards were successfully established in the Midlands. The temperature decline that followed continued for four centuries. The latter half of the 17th century included a period called the Little Ice Age when ice-fairs were held each year on the Thames. At the end of that century there were seven catastrophic years for farmers in the north of Scotland with harvests failing completely in four of them. The 18th and 19th centuries saw a repetition of this pattern, though the cold spells were not so extreme. Will the pattern repeat itself this century? Perhaps not, for the period from 1895 to 1940 was the warmest for some five centuries.

During adverse climatic periods only the better land will provide a livelihood. Some arable land is no longer worth cultivating and is allowed to revert to pasture. Land 'marginal' for pasture is abandoned and, below the tree line, undergoes secondary succession back to woodland. As the area under cultivation in Britain increased, so the area of marginal land going in and out of agricultural use increased. It has been suggested that the peak in ash (*Fraxinus excelsior*) pollen around 1100 A.D. reflects the amount of abandoned marginal land available for colonisation, rather than a more favourable climate for ash growth.

Of course, climate was only one factor affecting agricultural practice. The others we can group as sociological and they can be very complex indeed. Merton (1970) studied the woodlands of the Derbyshire limestone. He found that most of them were dominated by ash, and had originated since 1810.

The age-structure of the woods was investigated. One wood showed an age range of 50–60 years suggesting very slow establishment. Others had an age range of only ten years indicating rapid successful colonisation by ash. Hawthorn bushes which had been grazed when young, kept their distinctive form and provided a clue to the grazing intensity when secondary succession back to woodland was initiated. Seven different pathways for succession are postulated, depending not only on the abruptness of termination in grazing but also on the terrain and soil type. But what led to these marginal lands being abandoned? The various Enclosure Acts between 1764 and 1840 undoubtedly contributed. Radical changes in the communications of the region were also occurring. The old roads had kept to the limestone plateau. New roads, made to much higher standards, were for the first time following the valleys. Industry favoured sites beside these roads where there was also water. The tendency for people to move away from the plateau was accelerated by the shortage of capital for pasture improvement in the decade after the Napoleonic wars.

So far the history of British woodlands has been presented as one of progressive destruction since Neolithic times. As Man developed more permanent settlements and a more sophisticated social organisation, so his per capita use of woodland products grew while population numbers increased as well. It is fortunate in some ways that woodland products are heavy and expensive to transport while at the same time having low intrinsic value. Firewood supplies must be available locally and even timber is not worth carrying far unless cheap water transport is available. John Adair remarks of Scotland in the 17th century, "The Woods being much worn out and decayed. For, unless in parts where there could be no transport, hardly any remains of the great and goodly Forests that were of old, are to be seen." It seems almost a law of human nature that Man must virtually exhaust a natural resource before he will take action to conserve it.

In the Middle Ages iron smelting was still carried out in 'bloomeries' near woods where the charcoal was manufactured. In the 16th century the demand for iron products, particularly for guns for warships, became so great that the Wealden oakwoods were devastated. The first Act of conservation was passed in 1556 prohibiting the felling of wood in Sussex for the casting of guns. Other such Acts followed and by the turn of the century the iron smelters had moved the centre of their activity to Scotland. In 1609 an Act of the Scottish parliament aimed to control the making of 'Yrne with Wode' but it could not be enforced. At this time the larger towns consisted almost entirely of wooden dwellings. Within a short span in the 17th century there were catastrophic fires in London, Edinburgh and Glasgow. Timber had to be imported to rebuild them. The Government at last took positive action to restore its woodland resources. All estate owners were required to plant annually for ten years, one acre of woodland for every £250 of rental value, the plantations to be free of tax. No grazing was to be allowed in these plantations and trespassers were to pay a £5 fine or give ten days labour in lieu— very heavy penalties at the time. The same John Adair noted in 1703 that "the Inhabitants have planted of late such numbers of young trees of all Kinds that in a few Years there can be no want of timber for all uses." He was over-optimistic, of course. In 1812, a census of woodland in Scotland showed that only half a million acres of woodland remained. Some 400 thousand acres of plantations slightly redressed the balance.

Wars have always been periods when huge inroads were made into our remaining woodland resources and the Napoleonic wars were no exception. However, two developments reduced the pressures on woodland for a while. Coal replaced charcoal in the smelting of iron and steel replaced timber in the hulls of ocean going ships. The huge areas of woodland managed as coppice became progressively less profitable, though the tan bark market kept some going until the beginning of this century. Many of today's mature oakwoods have developed from coppice. Sometimes the coppice regrowth after the last exploitation was singled leaving the best coppice shoot on each *stool* to become a tree, making it difficult to find evidence of origin today. No mature tree that has arrived in this way has the size or the cleanness of bole of a tree grown direct from a seedling. Ruth Tittensor (1970) has described in some detail the history of coppice management in the extensive oakwoods of the Montrose estate on the east bank of Loch Lomond.

For 150 years now there has been little incentive to plant broad-leaf trees, and practically all the plantations of this period are of conifers and to a very large extent exotic conifers. Larch (*Larix decidua*) was very popular at the beginning of the period; later Douglas fir (*Pseudotsuga*) and Norway spruce (*Picea abies*) found favour, and in the vast expansion of planting since the last war Sitka spruce (*P. sitchensis*) has predominated. The greater part of this new forest estate was established on disforested land. True woodland flora will be slow to become re-established and of course all the young plantations in the pole stage (the building phase of the stand cycle) will have hardly any groundflora at all. Coniferous litter tends to be slower to decay and soil characteristics are likely to change in time where conifers have been planted on old hardwood ground.

This short account can do no more than indicate the kinds of historical factor that have played a part in determining the nature of our woodlands today.

Practical work—history

No set exercises are suggested here. Pollen analysis is extremely complicated and the specialist experience required in the identification of pollen grains and spores takes years to acquire.

In considering whether a full investigation of more recent historical factors would make a good project there are broadly three types of situation:

1. the wood is of recent origin.

2. the wood is known to have a long history and there is much published material about it.

3. the wood is suspected to have a long history but it does not appear to have been written about specifically.

The first offers very little scope, beyond discovering or inferring events relating to its origin and development.

The second offers the kind of project where the student is encouraged to discover for himself, on the ground, the evidence of past events. This could be linked with the interpretation of a certain amount of qualitative and quantitative data obtained from an enumeration of low intensity.

The third might make the most rewarding project though this depends on the amount of evidence that can be found. Old maps and lithographs, statistical accounts for the county, Forestry Commission census data, Estate records, novels set in the area and old travel books are some of the sources that could be used.

Historical records often give information of a general kind, or they refer to a specific wood using a name no longer recognised. Quite a lot of 'detective work' may be involved in determining how a well documented history applies to a particular locality. Boundaries change. Woods are added to or partially cleared. Keep an eye open for the remains of old walls, dykes and disused bridle paths. Hedges may provide clues. Pollard (1973) shows how hedges that were once woodland margins keep their high species diversity for a long time after the wood has been cleared. Hedges known to have been planted diversify very slowly and acquire a different assortment of species. Bits of charcoal in the topsoil may lead to the location of charcoal burning hearths and bloomeries where iron was smelted.

A drainage pattern may point to a previous period under agriculture. Exotic planting was a favoured method of providing cover for game birds. When a planted wood is old the original planting spacing will no longer be obvious, but a careful check of distances apart in a small area may provide the evidence needed. Evidence of past coppicing, multiple stems or the poor shape of singled stools, has been mentioned. Exotic tree species will have been planted. If they are found only in small groups they can perhaps be ignored in trying to establish the status of wood in which they occur. Some of the references below will provide further examples of the kind of inferences that can be drawn.

Anderson, M. L. 1967. *A History of Scottish Forestry,* Oliver and Boyd.

Fairbairn, W. 1972. Dalkeith Old Wood. *Scottish Forestry* 23 (1).

Merton, L. F. H. 1970. The history and status of woodlands of the Derbyshire Limestone, *J. Ecol.* 58 (3) 723–744.

Pennington, W. 1969. *The History of British Vegetation,* E.U.P.

Tansley, Sir A. G. 1939. *These British Isles and their Vegetation,* C.U.P.

Tittensor, R. M. 1970. History of Loch Lomond Oakwoods. *Scottish Forestry* 24 100–118.

5 Populations in the ecosystem

Each species population plays a part in one or more of the major processes that make up ecosystem metabolism. That role, the **ecological niche** of the species, is defined in answers to the following questions:

What kinds of energy sources does the species use?

We have already used several terms conveying this kind of information in a general way. Thus heterotrophs use organic matter and herbivores the live organic matter of plants. The exact energy source is indicated more precisely in terms such as *algivore, rhizophagous herbivore, seedling pathogen* or *pupal predator*.

Where is the function carried out?

Here we are concerned with the location in the ecosystem, though information about geographical distribution and the kinds of ecosystem a species inhabits may provide clues to the microhabitat preferences in the ecosystem. Ultimately location is defined by a number of critical environmental parameters as well as the general location of its food and it may therefore change with season and developmental stage.

When is the function carried out?

Each species tends to have a characteristic pattern of daily and seasonal activity. The long-eared owl (*Asio otus*) is a *nocturnal carnivore* in coniferous woods while the celandine (*Ficaria verna*) is a *vernal autotroph*.

How is it carried out?

The nightjar (*Caprimulgus europeaus*) may eat much the same assortment of insects as a spider but one takes them on the wing and the other traps them in a web.

49

Answers to these questions began to accumulate in the natural
history phase of ecology and then in increasing amount through the
work of autocologists combining field observation and experi-
mentation.

Tabulate all the relevant data about the niche of a well studied
species and it becomes obvious that the niche is unique. There can be
no other species which has exactly the same energy source, taken
under exactly the same environmental conditions, at the same time
and in the same way. However, if you identify a niche any less
precisely then more than one species may belong to it. Thus it is
possible to speak of *niche equivalence* between species filling the same
kind of role in different ecosystems or even between two or more
species that fulfil rather similar roles in the same ecosystem. Unfortunately, perhaps, the term niche has been used in other branches of
ecology without a functional connotation: a 'nesting niche' or a
'brackish water niche', for example, tell us only about the habitat
of a species, about the 'address' rather than the 'profession', as
Odum put it.

The first attempts to present a simplified picture of ecosystem
metabolism were based on food relationships. The primary energy
store, the biomass of autotrophs, is the first **trophic level**. It is consumed by herbivores which are therefore at the second trophic level.
Thus tree foliage, leaf-eating caterpillars, blue tits (*Parus caeruleus*)
and sparrow hawks (*Accipiter nisus*) comprise four trophic levels
linked in a **food-chain**. The food-chain, a valid and useful concept, is
an over-simplification for many purposes, for most animals vary their
diet and in turn are preyed upon by a variety of carnivores.

Classification of ecosystem biota by trophic levels has produced
some interesting generalisations of which the most important is that
productivity rapidly decreases at higher trophic levels. There cannot,
therefore, be many trophic levels. Four or five is the commonly
quoted number, though parasites and hyper-parasites are not
reckoned nor has a full analysis of decomposer food relationships
been attempted.

In practice few species can be tidily allocated to one trophic level.
The sparrow hawk, for example, feeds on small woodland birds but
these may be predominantly seed-eaters (trophic level 2) or predominantly insectivorous (trophic level 3). Think, too, where you
might place omnivores like ourselves. A better picture of food

relationships is conjured up by the term **food-web** which has no
connotations of specific levels.

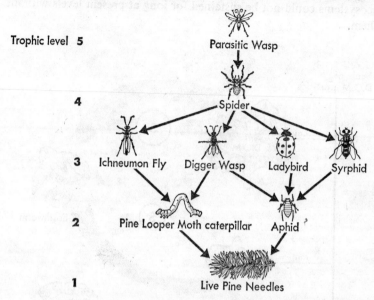

Trophic level **5**

4

3

2

1

Fig. 5.1. A simple food web (from: Richards, 1926).

Figure 5.1. shows a simple food-web in which it happens that
trophic levels are reasonably distinct. It is convenient sometimes to
distinguish between herbivore food-webs in which herbivores are the
primary consumers and decomposer food webs originating in the
primary consumers of D.O.M. There are innumerable links between
the two webs. All herbivores and carnivores contribute faecal
material to the D.O.M. store. Predators may find prey in both webs.

These analyses of organisation are useful but they do not yet shed
much light on the central processes in ecosystem metabolism, the
continuing biosynthesis by green plants which is dependent on radiant
energy and the return of vital nutrients through the biodegradation of
the material they produce. The former has been the subject of
intensive study at all levels up to the community and need not
concern us here. The latter involves many plant and animal populations. It might seem at first that animals are superfluous, for
heterotrophic plants, the **saprophytes**, are the main agents of D.O.M.
decomposition. However, even a brief consideration of what happens

in the decomposer food-web will make it clear that animals are deeply involved in its metabolism—that primary production in most ecosystems could not be sustained for long at present levels without them.

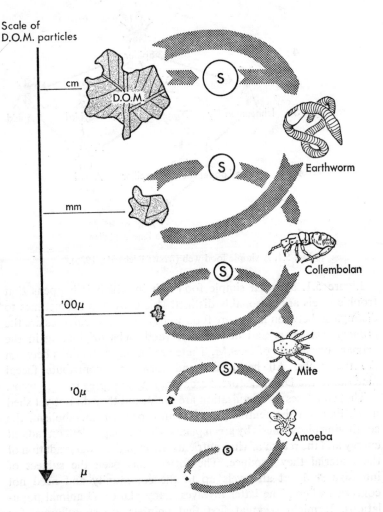

Scale of
D.O.M. particles

cm

mm

'00μ

'0μ

μ

D.O.M.

Earthworm

Collembolan

Mite

Amoeba

S = Saprophytic decomposition

Fig. 5.2. The Comminution Spiral.

Most of the larger animals (*macrofauna*) and many of intermediate size (mesofauna or meiofauna) subject the organic matter they ingest to chewing or grinding prior to digestion. *Comminution* enables digestion to take place more rapidly. Now animals tend to have much higher metabolic energy requirements than plants because they are mobile. A greater proportion of their assimilates is used up in respiration and a smaller proportion in forming new biomass. So it happens that faecal material egested tends to have a lower carbon/nitrogen ratio than material ingested (the C/N ratio is generally indicative of the proportion of carbohydrate to plant nutrient elements). Faeces are therefore a better substrate for saprophyte growth than the original organic matter.

The fact that, with minor exceptions, only saprophytes possess the enzymes capable of breaking down structural carbohydrates, particularly celluloses and lignins, leads us to speculate further about the integration of animal and plant roles in decomposition. Saprophytes release large quantities of assimilable carbohydrates: but their metabolic energy requirements are small. They find themselves with a surplus of carbohydrate for the formation of new saprophyte biomass and a corresponding deficit of other nutrient elements. During active growth saprophytes actually withdraw nutrient elements from the soil solution, competing directly with autotrophs. It has been known for trees mulched with sawdust to show nutrient deficiency symptoms in their leaves: the saprophyte activity stimulated by the addition of a new energy source, itself poor in nutrients, has depleted the nutrients that would have otherwise been available to the trees. Studies of *mycorrhizae*, fungi associated with roots, suggest that fungi are likely to be more efficient than higher plants at taking up nutrient elements from dilute solutions: the loss of carbohydrate reserves from the roots is more than compensated by the increased efficiency of nutrient uptake. This hypothesis is supported by the absence of mycorrhizae from very fertile soils where their formation would give the plant no competitive advantage and from very infertile soils where there is virtually no surplus carbohydrate available in roots to support the fungus.

Saprophytes, then, accomplish a major part of organic matter decomposition because only they can degrade structural carbohydrates, but, like autotrophs, they make nutrient element demands on the system. Decomposition is not synonymous with **mineralization**, a term used to indicate the processes whereby mineral nutrients are released into the mineral nutrient pool and become available again for uptake by plants. Very small amounts of nutrient minerals are leached by rainfall from all living organisms and all D.O.M. The most likely source of major contributions to the nutrient pool is the faecal material of animals. Think not only of deer dung or owl pellets but also of worm-casts and the faeces of the myriad other decomposer web animals. Primary consumers in the decomposer web tend to be classed as **saprovores**. Earthworms certainly fall in this category. But many non-predatory mites are known to feed on fungal spores and hyphae: protozoa feed largely on saprophytic bacteria. Indeed it seems a tenable hypothesis that even true saprovores have as their main energy source, the saprophyte biomass on or in the D.O.M. they consume.

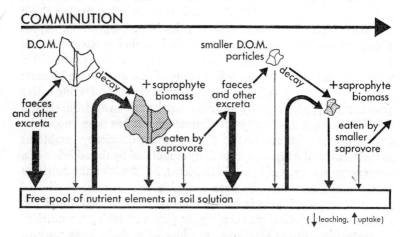

Fig. 5.3. A tentative scheme integrating the roles of saprophytes and saprovores in the decomposition process and the release of nutrient elements.

The speculation in the above arguments is revealing. Border areas between established disciplines have tended to be neglected—here a border area between botanical and zoological ecology. Animal activity, it would appear, largely determines the rate of mineralization; directly through the egestion of readily leached nutrient-enriched faecal material and indirectly through the increase in saprophyte activity made possible by comminution and the lowering of the C/N ratio of D.O.M.

Figure 5.4 overleaf is yet another diagram of ecosystem metabolism which should be compared with similar diagrams in other ecosystem ecology texts. Note the inclusion of the *comminution spiral* and the faeces subdivision of the D.O.M. store providing the main source of replenishment of mineral elements in the soil solution. Herbivores include saprovores, which simplifies the diagram but raises another problem. By consuming biomass herbivores reduce the productivity potential of their food plants. Should herbivores become super-abundant the plant biomass may be depleted to a point where primary productivity is too low to support the ecosystem in all its complexity. It has often been suggested that saprovores do not offer such a threat to ecosystem metabolism because the consumption of all available D.O.M. does not put at risk the future production of D.O.M. Note, however, that the 'saprovores are herbivores' hypo-thesis implies that saprovores will reduce the productivity potential of saprophytes. We do not have the data to determine whether this adverse effect on decomposition is outweighed by the favourable effect of comminution and lowering of the C/N ratio of D.O.M. on saprophyte productivity. Finally, we may note that other animals, the carnivores and parasites, tend to reduce the numbers of both herbivores and saprovores. The interaction between predator and prey population numbers is taken up in Chapter 8.

Three major processes are usually shown as activating the Mineral Cycle; *uptake*, *turnover*, and *decomposition*. Of these, decomposition is the slowest and most likely to limit the overall rate of cycling. But we have seen that any decomposition resulting in increased saprophyte biomass must lock away additional nutrient capital. Saprovores play a critical role in its release. Without animals cycling rates would be very much slower.

An increased rate of nutrient cycling compensates for lack of nutrient capital in the ecosystem, which is why high productivities are found in tropical rain forest despite the fact that they occur mainly on rather infertile soils. We have noted that there one cycle may be completed in four to six months, a speed largely made pos-sible by the activity of termites. In zones where winter cold or dry-season drought prevent continuous primary production, uptake is concentrated at the beginning of the growing season while turnover and decomposition peaks of necessity follow in an annual cycle.

During progressive succession we have seen that biomass grows and turnover increases, an indication that the proportion of nutrient

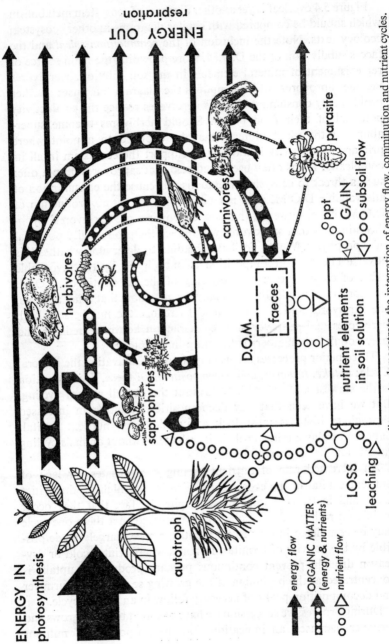

Fig. 5.4. Simplified ecosystem process diagram to demonstrate the integration of energy flow, comminution and nutrient cycles.

capital cycling is also increasing. Succession leads to an increase in both structural and floristic diversity which provides for a potential increase in the number of niches in the herbivore and decomposer food webs. Full diversification of the decomposer web comes last and with it the maximizing of cycle rates.

These generalizations about cycle rates seem to be soundly based though we have surprisingly few estimates even for individual nutrient elements. We can analyse the various nutrient stores in the ecosystem and see how they change with time: but this is not enough. Rainfall leaches nutrient elements differentially from everything it wets, biomass, D.O.M. or mineral particles, while at the same time introducing small amounts. Figure 5.5 illustrates the results obtained in a study of this kind in a sessile oak wood.

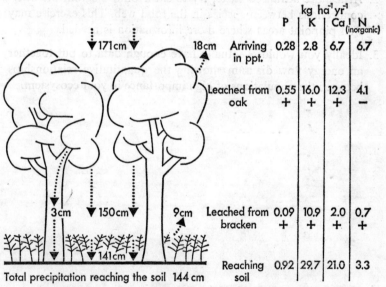

kg ha⁻¹ yr⁻¹

		P	K	Ca	N (inorganic)
▼171cm▼	18cm	Arriving in ppt.	0.28	2.8	6.7	6.7
		Leached from oak	0.55 +	16.0 +	12.3 +	4.1 −
3cm ▼150cm▼	9cm	Leached from bracken	0.09 +	10.9 +	2.0 +	0.7 +
▼141cm▼		Reaching soil	0.92	29.7	21.0	3.3

Total precipitation reaching the soil 144 cm

Fig. 5.5. Partial nutrient budget in a sessile oak wood (from: Carlisle *et al.*, 1967).

There are other imports, in ground water flowing in to illuvial sites, through immigration of biota, D.O.M. blown in, or the weathering of soil parent material. There is an equivalent series of losses, in ground water outflow, particularly from eluvial sites, emigration, soil erosion etc. A complete nutrient element budget for the ecosystem appears to be a necessary preliminary to reliable estimation of cycle rates.

Practical work-populations in the ecosystem

The kind of work appropriate here requires a lot of information about the component species in the ecosystem of the kind described in the next three chapters. Plant data is likely to become available sooner than animal population data.

1. Describe as accurately as you can the ecological niche of the groundflora species of high constancy. This will involve a study of seasonal changes in the populations (phenology) as well as the soil and other characteristics which appear to determine their location.

2. At a more advanced stage, try to show on a diagram the interconnections between species in the food web. This exercise may well pinpoint areas where more information is needed.

3. Ideally you would eventually have enough data to put together an energy flow diagram showing the quantitative relationships between the populations of high importance in your ecosystem.

6 Assessing the relative importance of species populations

Soil fauna

All species active in an ecosystem are potentially important. There is usually an apparent superfluity of species. Part of this diversity can be seen to allow the adaptation of ecosystem metabolism to changing environmental conditions in either the short or the long term. It must always be remembered that an ecosystem investigation at one point in time leads to an analysis only valid for that point in time. Where there are seasonal patterns of activity, these patterns must be studied, as must variations from year to year. The general level of the species population will determine its continuing contribution to ecosystem functioning and it is this general level which should be used to determine the relative importance of species.

To discover the species of high relative importance we need to do three things:

1. identify at least the more abundant and the larger biota present.

2. estimate population size and structure.

3. estimate the contribution of individuals at various stages of development to overall population activity.

Success in estimation of population numbers often depends on a detailed knowledge of niche and behaviour. Techniques tend to be tailored to taxonomic groups and trophic levels—some to a single species at a time. Estimations of activity must usually be based, rather precariously, on measurement of physiological parameters in the laboratory. Examples of the kinds of techniques that have been evolved are dealt with under sub-system headings overleaf.

C

I Primary producers

A The tree stratum

In Britain identification problems are minimal because of the small number of species established in woodland. Single species dominance in most woods further simplifies the problem. Population size and structure are determined by enumeration (see Chapter 1). Fortunately the growth pattern of trees is very systematic. There tend to be good correlations between the many measurements that can be made. For many purposes it suffices to record only the girth of each stem at breast height (1·3 m).

Enumeration details the species, their numbers and their girths in each sample unit (a strip or preferably quite a short sub-division of each strip). You will notice that when there are many trees in a sample unit the individuals tend to be small—and vice versa. A measure which integrates size and number is **Basal Area,** the sum of the cross-sectional areas of all tree stems (at breast height) in an area. Conversion from girth (or diameter) to basal area assumes that departures from circular section will be small and this is usually true. *Relative Basal Area* is thus a good indicator of the importance of particular tree species components:

$$\frac{\text{Relative Basal Area \% of species A}}{} = \frac{\Sigma \text{ Basal area of trees of species A} \times 100}{\Sigma \text{ B.A. of all species}}$$

When this value is 90 or over as it is likely to be in a successful single-species plantation (any other tree species having established themselves naturally) then the statistics for species A will approximate to those for the tree component as a whole. This applies even when the minority species have rather different growth patterns. Thus in Table 1.1 about a quarter of the trees classified as 'suppressed' were in fact birch saplings: the fact that birch grows in rather a different way from pine is not likely to affect whole crop parameters to a significant extent.

The next step is to discover the relationship between biomass and girth for the major species. We will expect it to be curvilinear; for weight is proportional to volume and volume is a cubic function of girth or diameter. We will also expect correlation to be good, especially in immature crops and for conifers with their more regular growth pattern. To determine the line of best fit for data lying about a curve is much more difficult than for data lying about a straight line. Ecosystem analysts in these situations tend to investigate various transformations of the data which might change the relationship to a linear one. Here a logarithmic transformation does the trick:

$$\log \text{ dry wt. (kg)} = a + b \log \text{ girth (cm)}$$

This straight line can be fitted with reasonable precision by eye (remember it must pass through the point corresponding to the mean log dry weight and the mean log girth) or very precisely by calculating the *regression* of log dry weight on log girth. Most textbooks on statistics will describe the computations required.

Ideally a large number of trees covering the size range in the plot are felled and weighed. In practice a number as small as eight has given reasonable results. These trees should come from the plot surround if possible. Sometimes data from thinned trees may be available, though thinnings are not often representative of the stand as a whole. At Brandon Heath Ovington (1957) calculated the mean height and girth for each of his plots and then felled one tree with approximately these measurements in each. It can easily be demonstrated that the mean girth tree is likely to give an underestimate of the mean dry weight, (see Fig. 3.7). The mean basal area tree would have been better and could have been computed from the data he had. The mean stem volume tree would have been best but a great deal more field work would have been required to calculate individual tree volumes. A rather tedious compromise is available if forester's volume tables have been published: these give volume estimates against girth, (Forestry Commission).

Having calculated the regression of dry weight on girth, work out the girth of the mean basal area tree in your plot and insert this value in the regression equation to obtain its dry weight. This weight multiplied by the number of trees in the plot is your estimate of biomass.

The activity of the tree population manifests itself in the formation of new tissues (production) and in respiration (energy for maintenance). It is usual to ignore respiration and deal simply with **net production** because of the difficulty in estimating respiration while photosynthesis is occurring. Nevertheless it is important to remember that respiration accounts for an increasing proportion of gross photosynthetic production as the tree ages. Net production may go into temporary structures such as leaves, flowers or bracts (when it is called *turnover*) or into more permanent parts of the root stem and branch system (when it constitutes *biomass increase*).

Estimation of biomass increase (or conceivably decrease) requires two enumerations separated by a long enough period for measurable growth to have occurred, often one or two years. If errors creep into the estimate of biomass, they are likely to be similar for both enumerations and the difference, biomass increase, may well be estimated more precisely than the biomass itself.

It is possible to make a crude estimate of biomass increase with only one enumeration if the growth rings in the stem are clearly countable. Small cores, two or three centimetres long, can be taken at breast height with a Pressler borer. Measurement of mean annual ring width allows estimation of growth in previous years (assuming that bark thickness has not changed significantly). The regression equation then gives the biomass estimate in the same way as before. The method cannot take into account any mortality that may have occurred.

Turnover may be estimated in part by setting litter traps and, at intervals, collecting the material falling into them. There is a continuous minute loss as decay proceeds. Possibly more important, leaching will occur during wet weather. However there is little dry weight loss and little litter fall in the winter months; two collections in January and April will suffice for many purposes. From May to October (in Britain) collections should be more frequent, and not less than monthly. The simplest type of trap—a tray with wooden sides and a fine nylon mesh bottom—is virtually as effective as more expensive funnel or bin traps. It should be supported in a horizontal position just above ground level by four small pegs. The trap loses fine particles of organic matter and pollen to the extent of 10–15 grams dry weight per square metre per year. Random location of traps is indicated if you wish to estimate the precision of your mean litter fall figure.

Collections are oven-dried and weighed. They may be sorted out into as many categories of component as seem likely to give data of value. It will be found that some components are estimated with much greater precision than others. Any items that are light and widely dispersed tend to be more uniformly distributed between traps, giving higher precision in their estimation. Thus leaves, bracts, flower parts and winged seeds will have better estimates than twigs, bark and cones.

TABLE 6.1

Precision of yearly litter component estimates in a pine plantation.

Component	Dry wt per trap, g	Confidence limits (95%)	Dry wt. ha^{-1} tonnes
Needles	80·0	±7·7%	3·20
Cones	18·5	±30%	0·74
Twigs and bark	9·3	±24%	0·37
Miscellanea	10·0		0·40
Total	117·8	±10%	4·71

The number of trays and their size can be varied, though for statistical reasons it is advisable to have at least twenty. The figures in Table 6.1 above are based on twenty tray traps and twenty funnel traps randomly located in a quarter hectare plot. Each trap was about $\frac{1}{4}$ m^2 and the sampling intensity was thus about 0.4%.

The remaining part of the above ground turnover for which trees are responsible comprises whole trees and branches that die. Decay begins on the tree; very very slowly parts disintegrate and fall. In windy localities dead trees may be blown over and dead branches snapped off: exceptionally, live trees and live branches will fall. Such individually large items can be weighed in the plot but there can be little confidence that data collected in a limited area over a few years will be representative of wider areas.

Standing dead trees can be enumerated and sampled in the same way as live trees. The regression of their dry weight on girth will be very similar to that for live trees and total above ground D.O.M. in standing dead trees can be calculated in the same way. Recruitment and losses can be calculated by comparing successive enumerations of live trees.

Dead branches attached to the tree are normally assessed at the same time as the biomass when sample trees are dismantled for dry weight estimation. A regression of dead branch dry weight on live tree girth can be calculated, providing for a crude estimate of the yearly recruitment/loss balance.

The investigation of what is going on under the ground presents much more intractable problems. The situation is analogous to that above ground. There is a massive framework of large woody roots and a mass of short-lived fine rootlets, which, like the leaves above, provide the bulk of the turnover. Fine rootlets can be estimated on an area basis by taking a series of soil monoliths back to the laboratory and carefully washing out the roots. The smallest of these rootlets (the upper limit is commonly taken as half a millimetre diameter) provide an estimate of fine root turnover on the assumption that they live for two years (Orlov 1955). Exceptionally the destruction of the ecosystem involved in excavating the large roots may be justified. The roots of the trees taken as samples for the estimate of above ground biomass would then be washed clean of soil with hoses, dismantled and weighed. Growth ring analysis, as described for the stem, would allow an estimate of biomass increase. To date there have been few estimates of root biomass and even fewer estimates of root productivity though the position should improve tremendously as the results of the International Biological Programme investigations are published. Orlov (1955) concluded that the ratio of turnover to biomass in the underground parts of a tree is similar to that above ground.

Figure 6.1 is a flow diagram constructed to assist students in a project aiming to estimate tree productivity in a pine plantation. Note the use of outside data to cross-check project data and to interpolate values for items not investigated.

Some conifer species exhibit very systematic growth patterns, readily linked with age. First year, second year, and older twigs can be separated and their dry weight estimated. The oldest part of the branch can be analysed through its growth rings. Furthermore whorls of branches are produced regularly, one a year, in temperate climates. This kind of situation allows other approaches to the estimation of crown productivity. Baskerville (1965) working in a balsam fir (*Abies balsamea*) stand was able to establish a trend of mean whorl dry weight with age. As the trees grow the whole canopy moves upwards, death of branches at the base of the live crown being roughly balanced by biomass increase above. Ovington's data (1957) from Brandon Heath pine plantations show clearly that crown biomass per unit area remains substantially the same for several decades of the building phase and early mature phase. If we assume no change over a short period, then the curve of whorl dry weight increase and then decrease with age will provide an estimate of branch turnover. The regularity of crown structure also assists in the *stratification* of the crown when subsampling to determine dry weight (see Fig. 6.2).

B Shrubs and groundflora

Here we are dealing with an assemblage of species with many different life forms each requiring a separate technique in biomass and productivity estimation. Large shrubs may be treated like trees if they have a single stem which is reasonably cylindrical. If the stems are gnarled and twisted, or if there are multiple stems, then some ingenuity may be required to devise a suitable parameter for measurement which does not take up a disproportionate amount of time, yet correlates well with dry weight. Height alone will probably show poor correlation with dry weight: 'height times diameter' is sometimes well correlated.

Dwarf shrubs such as heather and bilberry (*Vaccinium myrtillus*) are small enough to be harvest sampled in the plot at low intensity without doing lasting damage. Turnover can be estimated with miniature litter traps in pure stands; otherwise it must be estimated, with biomass increase, by growth ring and shoot analysis. Scrambling shrubs such as brambles (*Rubus fruticosus* agg.) require detailed investigation of their structure through the year, before realistic estimates of their productivity are likely to be obtained. This observation applies to many groundflora species and the assumption here is that you will first make a subjective assessment of the likely importance of each species to the ecosystem and to your particular investigation. When the importance rating is high individual species studies are indicated. Of course someone else may have looked at the problem already. Blackman and Rutter (1946) studied bluebell (*Endymion non-scriptus*) populations in a wood near Oxford. They showed that early development depended on food reserves in the bulb and it was mid-April before biomass exceeded initial bulb weight. In the following seven weeks biomass grew to three times the initial bulb weight. Subsequently the above ground parts died back leaving new bulbs with a weight 1·6 times the initial bulb weight. Presumably the population was expanding that year or there was appreciable mortality of bulbs. This study tells us that early June would be the best time for an assessment of maximum biomass, that turnover is of the order of 50% of that maximum biomass, that production was about twice the initial bulb dry weight.

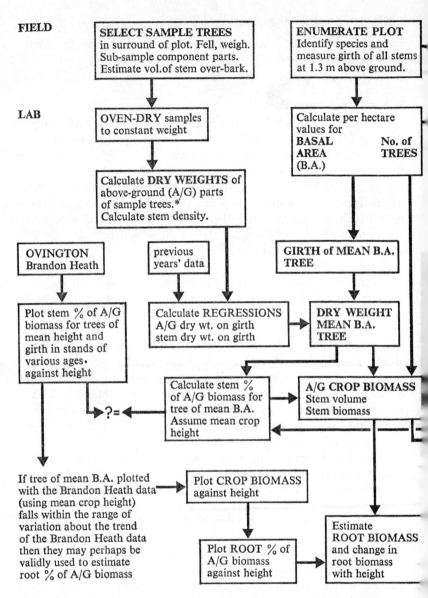

FIELD

SELECT SAMPLE TREES
in surround of plot. Fell, weigh.
Sub-sample component parts.
Estimate vol. of stem over-bark.

ENUMERATE PLOT
Identify species and
measure girth of all stems
at 1.3 m above ground.

LAB

OVEN-DRY samples
to constant weight

Calculate per hectare
values for
BASAL No. of
AREA **TREES**
(B.A.)

Calculate **DRY WEIGHTS** of
above-ground (A/G) parts
of sample trees.*
Calculate stem density.

OVINGTON
Brandon Heath

previous
years' data

GIRTH of MEAN B.A.
TREE

Plot stem % of A/G
biomass for trees of
mean height and
girth in stands of
various ages,
against height

Calculate REGRESSIONS
A/G dry wt. on girth
stem dry wt. on girth

**DRY WEIGHT
MEAN B.A.
TREE**

?=

Calculate stem %
of A/G biomass for
tree of mean B.A.
Assume mean crop
height

A/G CROP BIOMASS
Stem volume
Stem biomass

If tree of mean B.A. plotted
with the Brandon Heath data
(using mean crop height)
falls within the range of
variation about the trend
of the Brandon Heath data
then they may perhaps be
validly used to estimate
root % of A/G biomass

Plot CROP BIOMASS
against height

Plot ROOT % of
A/G biomass
against height

Estimate
ROOT BIOMASS
and change in
root biomass
with height

*Dead branches attached to the stem are not considered part of the biomass.
Their dry weight will allow a crude estimate of branch turnover on live trees.

Fig. 6.1 Flow diagram for Productivity Estimation in a pine wood.

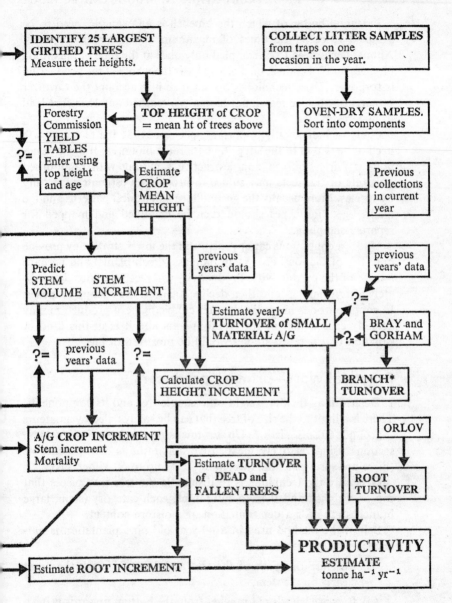

?= indicates that a cross-check can be made. Serious discrepancies suggest errors in technique *or* incompatibility of data.

Perennial herbs, of which the bluebell is an example, need to be sampled twice at the times of maximum and minimum biomass. Annual herbs need to be sampled only once at the time of maximum biomass and, of course, on a yearly basis the whole of the production is turnover. These techniques do not take into account the turnover of parts lost before maximum biomass is reached and therefore lead to underestimates.

Study of individually important species outside the plot area will need to be tied in to the situation within the plot area. If the ground-flora is fairly uniform non-destructive sampling with randomised quadrats or transects may suffice to provide an estimate of density by species. More usually the groundflora will exhibit patterns and the main vegetation types should then be identified and mapped for separate sampling.

Under a continuous canopy, plants in the lower strata may provide only 1% of the total ecosystem biomass (green plants) and perhaps 5–10% of the productivity. Productivity will receive a boost immediately after thinnings and then decline again as the canopy re-closes. In coppice woods immediately after coppicing herb production may actually exceed that of the woody plants and it is in this kind of situation that detailed studies would be most rewarding.

Practical work—primary producers

1. Determining the dry weight of a sample tree and its components. There is a limit to the size of tree that can be safely felled by amateurs —around 60 cms girth.* Up to this size the weighing of stem sections using relatively inexpensive apparatus is also practicable (there are spring balances weighing items up to 50 kilograms).

Generally the technique is to dismantle the tree into pieces that can be weighed and then to sub-sample each category (stem, large branches etc.) for a determination of moisture content.

Below, the method used in a 40-year-old pine plantation is outlined:

(i) Measure the girth of the tree at 1·3 m and mark the point clearly on the stem.

(ii) Remove whorls of branches from the bottom upwards using a pruning saw or stout long-handled secateurs. Separate the

* If in any doubt as to competence in felling a tree safely, ask for expert help.

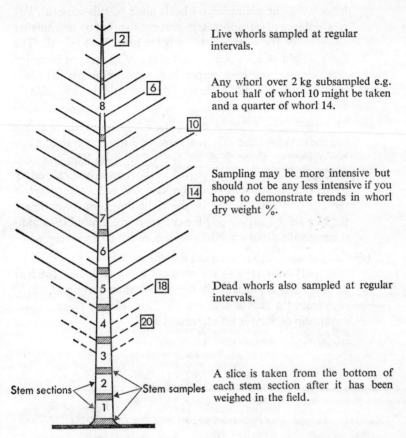

Live whorls sampled at regular intervals.

Any whorl over 2 kg subsampled e.g. about half of whorl 10 might be taken and a quarter of whorl 14.

Sampling may be more intensive but should not be any less intensive if you hope to demonstrate trends in whorl dry weight %.

Dead whorls also sampled at regular intervals.

A slice is taken from the bottom of each stem section after it has been weighed in the field.

Fig. 6.2. Dissection of a tree for the estimation of moisture content.
Table 6.2 and 6.3 show the kind of data that may be obtained.

dead and live branches in each whorl and weigh them in a polythene sheet. An extendable aluminium ladder is useful so that whorl removal can be done well into the live crown, and the tree (when cut) can be lowered down without damage to the remaining live branches.

(iii) Depending on the number of live whorls, 3–5 whorls at regular intervals down the crown are selected for dry weight estimation (see Fig. 6.2). Small whorls are brought back to the laboratory complete: large whorls must be sub-sampled. For convenience of transport large branches are cut up into smaller pieces after the needle-bearing twigs have been cut off. The fresh weight of each component must, of course, be recorded meticulously and each paper bag must be clearly labelled: polythene bags with separate labels should be available in case the weather is wet.

(*iv*) The tree is now felled. A cut is first made about 30 cm above ground level on the side you intend to lower the tree. Then begin cutting through from the other side. The tree will tend to remain upright supported by the crowns of surrounding trees. Gentle pressure from the saw side will help complete the cut and tilt the tree in the right direction. Now lower the tree gently, supporting the crown above ground level until the remaining live whorls have been removed and weighed.

(*v*) Cut the stem into convenient lengths for handling. A band of wide polythene tape at the crosscut point will limit the loss of bark: this is particularly important when cutting a sample slice from the end of each stem section. Finally clear around the stump and cut it off at ground level.

TABLE 6.2

Estimation of stem dry weight.

Section No.	Length m	Section fresh wt. kg	Sample fresh wt. g	Sample dry wt. g	Dry wt. %	Estimated section dry wt. %	Section dry wt. kg.
Leader	1·22	0·27	53	22·5	42·5*	41·25	0·111
7	1·71	2·8	55	22·0	40·0	41·25	1·155
6	2·00	11·8	350	184	52·6*	46·3	5·463
5	1·00	9·5	285	123	43·2	47·9	4·551
4	1·00	11·0	320	144	45·0	44·1	4·851
3	1·00	14·0	410	198	48·3	46·65	6·531
2	1·00	16·5	523	266	50·9	49·6	8·184
1	0·90	20·6	1170	526	45·0†	47·95	9·878
TOTALS	9·83	86·47					40·724

* These samples coincided with branch whorl positions and contained denser branch wood.
† It is normal for wood at the base of the stem to be less dense (Root wood is less dense).

Figure 6.2 and Tables 6.2 and 6.3 show how the work can be organised and the data set out to determine the dry weight of the above ground parts of a pine tree.

TABLE 6.3

Estimation of crown dry weight.

Whorl no.	Fresh wt. g	Sample fr. wt. g	Cones (†) g	Vege- tative parts g	Dry wt./Fr. wt. % Cones	Veg. pts	Whorl sample dry wt. Cones	Veg. pts	Total
1	136								
2	340	340	63	277	50·3	38·0	29·9	105·3	135·2
3	1 020								
4	1 120								
5	2 370								
6	1 845	1 845	129	1 716	51·2	40·0	66·0	686·4	752·4
7	3 120								
8	2 240								
9	3 370								
10	1 257	1 257	14	1 243	50·7	40·6	7·1	504·7	511·8
11	1 640								
12	370								
13	90								
*14	—								
15	7 700	1 974	6	1 968	51·7	43·4	3·1	854·1	857·2
16	3 970								
17	1 920								
Totals	32 508	5 210	212	4 998			106·1	2 150·5	2 256·6

To calculate crown dry weight

crown dry wt.

Method 1. Assume sample representative $\dfrac{32\,508 \times 2\,256\cdot6}{5\,210}$ = 14 080 g

Method 2. Assume whorl samples representative of part of the crown e.g. Wh. 6 represents Wh. 5–7 and half Wh. 4 and 8. = 13 593 g

Method 3. Assume cone dry wt. % constant. Graphs cone fresh wt. % of whorls. Graphs dry wt. % of vegetative parts. Estimates inter-sample whorls dry wt. using graph dry wt. %. = 13 647 g

Method 3 is recommended if graphical trends can be interpreted with confidence. Otherwise, Method 2 is preferable.

* Leader lost 14 years ago: whorl 14 is represented by a few small dead twigs.
† Cones treated separately because moisture content lower and did not change with position in the crown.

Note that for this particular tree cones have been treated separately: their moisture content is more or less constant while that of the vegetative parts shows a steadily increasing moisture content towards the top of the tree: furthermore the cones are most abundant on whorls 4–8. In Table 6.2 note that moisture content is significantly lower where branches join the stem. A method of ensuring that this denser timber received representation proportional to its extent in the stem would be a desirable refinement.

2. Estimating above ground biomass increase in the tree stratum.

   (i) Enumerate a plot in successive years at approximately the same time of year and preferably between October and May (outside the growing season).

  (ii) Calculate the mean basal area tree from all those trees alive at the second enumeration. Trees that have died go to turnover. The data will allow you to check that the girth of the mean basal area tree is indeed greater than the mean girth. If Forestry Commission volume tables are available the girth of the mean volume tree may be calculated and this again will be greater than the girth of the mean basal area tree, particularly if there is a wide range of size classes present.

 (iii) If your wood is dominated by one of the species dealt with in the Forestry Commission management tables (F.C. Booklet No. 34) a check of your estimate of basal area against their prediction may be made. It requires only that you measure top height and that you know the age of your wood. You can then enter the appropriate yield class table. (See also Chapter 9).

 (iv) Use your sample tree dry weight data to calculate
    (a) the regression of dry weight on girth
    (b) the dry weight of the mean basal area tree in each year.
    The difference multiplied by the number of surviving trees gives the biomass increase.

Generally you will find that while as few as eight sample trees may give you good correlation $(r+0.99)$ between above ground dry weight and girth, more trees are needed before the various tree components can be estimated with the same precision. One reason is that the proportions of stem, root and crown biomass vary between dominants, sub-dominant and suppressed trees, as well as with differing density in various parts of the stand.

Ovington's (1962) collation of woodland biomass data is a useful source of material against which to check your own estimates.

3. Rough estimation of groundflora biomass and turnover.

(*i*) Run several transects one metre wide across the plot, noting for each species of vascular plant the density and range of size of individuals. This should be done in mid-summer when most species will be close to their maximum biomass.

(*ii*) If mosses are abundant, map the mossy patches within the transects and sub-sample them separately on an area basis. A convenient method is to use a soil-corer to take some twenty samples each of about 200 cm^2, sorting out the moss from the litter in the laboratory.

(*iii*) Outside the plot look for specimens of the more abundant species within the size range observed. Excavate their root systems carefully and take them back for dry weight esti-mation.

(*iv*) To estimate productivity divide the species into dwarf shrubs, perennial herbs, mosses and annual herbs: then assuming the following relationship between maximum biomass and productivity make the necessary calculations:

(*a*) dwarf shrubs—30% biomass

(*b*) perennial herbs—50% biomass

(*c*) mosses—80% biomass

(*d*) annual herbs—100% biomass.

Assessing the relative importance of species populations

II The decomposers

This group presents some of the most intractable problems. Dead plant material very often starts to decompose while still attached to the living plant. Indeed some of the organisms that begin the decomposition processes in leaves have entered them as parasites before abscission occurs (e.g. the pine needle-cast fungus, *Lophodermium pinastri*): then, when the leaf falls, they continue their activity as saprophytes. There are many specialist wood-rotting fungi which hollow out the stems of living trees; they gain entry through wounds which expose the secondary xylem. But, decomposition takes place predominantly in the litter and the soil. Here there are several times the number of species found in the rest of the ecosystem; and many of them are micro-organisms belonging to obscure groups which are difficult to isolate and identify. We cannot yet describe the ecological niche of many of the species we identify which means we have a very incomplete knowledge of processes in the decomposer web. We still tend to refer to them all as 'decomposers' although many are predators and phytophagous animals (herbivores).

With little knowledge of many component population subsystems, a synthesis approach is out of the question. Yet if ecosystem analysis is to be meaningful we must attempt to estimate the activity going on in the decomposer web. Fortunately there are indirect methods. The first requires the monitoring of the amount of litter present on the woodland floor over a period of years, combined with estimations of the litter falling to the ground. If the litter layer stays the same, year in, year out, then decomposition must be equivalent to the litter fall. If litter is accumulating then decomposition is less. This method ignores possible soil organic matter changes. The second method seems, at first sight, more likely to give the kind of data we want. It involves the measurement of the carbon dioxide production of small areas of litter covered soil. There are many technical difficulties: the enclosure of small areas and the extraction of carbon dioxide from the air will almost certainly create abnormal conditions for the respiration of some soil and litter organisms. Then there is the question; "How representative is the small sample area?". This has to be considered in both the spatial and the time dimension: there are known to be diurnal and seasonal changes in activity patterns. Finally the method requires that the contribution of root respiration to carbon dioxide production be estimated. As you will appreciate from what we said about estimating root productivity (see Chapter 6), such estimates are likely to be very imprecise. (Anderson (1973) has made a recent assessment of this approach.)

Research data that may throw light on decomposition processes come from autecological studies of particular taxonomic groups and from the investigation of what happens to particular pieces of D.O.M. Fallen pieces of litter can be marked or put in nylon-mesh bags and replaced on the ground. From a large number of such items, samples are recovered periodically and note taken of dry weight, chemical composition, and micro-organisms present. There are practical problems. Only a wide mesh bag will allow all kinds of decomposer organisms access: yet a wide-mesh bag will also allow bits of organic matter to fall out when decay is well advanced. Figure 7.1 provides an example of results obtained in a study of this kind.

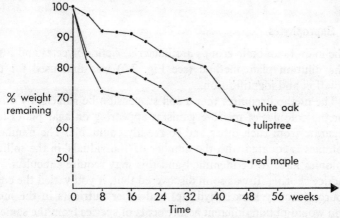

Fig. 7.1. Weight loss with time of leaves of red maple, tulip tree and white oak enclosed in open-mesh nylon bags in the litter (from: Thomas, 1970).

Later stages of decomposition cannot be followed by this method. Small pieces of comminuted D.O.M. are not identifiable and their decomposition is taking place below the surface litter where placement and recovery would disturb the microhabitat.

When the litter is abundant it is possible to distinguish three separate layers above the mineral soil—an upper *L layer* of freshly fallen material, an intermediate *F layer* where much breakdown has occurred but pieces of tissue can still be recognised and a lower *H layer* consisting of finely particulate organic matter. The fine particles of the H layer are small enough to be washed down into the top (A1) horizon of the soil where they undergo further decomposition. Investigation of chemical changes in D.O.M. from L layer to the A1 soil horizon seemed a promising line of enquiry at one time; but, as yet, such analyses have not given much help in the elucidation of the way in which decomposers operate. In a thick layer of the kind described above the organic or humus layer is said to be in the **mor** form characteristic of most conifer woods. This situation is contrasted with that in most broad-leaved woodland where little surface litter accumulates and the humus is said to be the **mull** form. Mull humus is thus associated with brown forest soils and mor humus with *podsolic* soils.

Autecological studies of plant decomposers (saprophytes) and animals usually classed as saprovores (consumers of D.O.M.) are dealt with separately below.

A Saprophytes

The main taxonomic groups are Bacteria, Actinomycetes and Fungi. The dilution plate method (see Fig. 7.2) was first used for their isolation and identification.

The method requires that a soil suspension be sufficiently diluted for the colonies of micro-organisms appearing on agar plates to be separate from each other and so readily countable. The number of colonies is equated with the number of 'individuals' in the soil. The colonies have to be identified and this may require subculturing to induce sporing. It was soon discovered that, if you varied the energy source (usually a carbohydrate) and other nutrients in the media, you would obtain different assortments of species from the same soil suspension. A range of media would obtain better representation of the species present but there is a limit to the amount of replication

DECOMPOSERS

that is practicable. In the many long lists of fungi identified from soils, Basidiomycetes, the common toadstool and bracket fungi of our woods, are conspicuous by their rarity. Their activity, it seems, must be concentrated in the L and F layers.

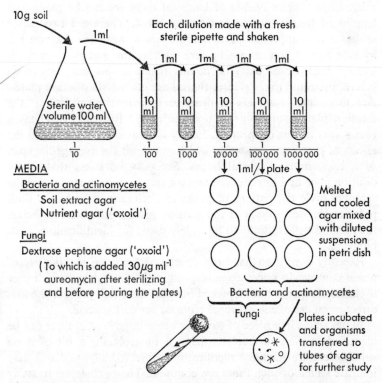

Fig. 7.2. Dilution plate method (from: Jackson and Raw, 1966).

The dilution plate technique seems unsatisfactory in many ways. It assesses accurately the numbers of those bacterial species capable of developing on the media provided; but the fungal count is much less meaningful. A sporing mycelium may be represented in the count by hundreds of colonies on the dilution plates, while a non-sporing mycelium may not be represented at all. It has been estimated that only some ten per cent of hyphal fragments in soil suspensions actually form colonies on agar. The method fails to distinguish between true members of the soil microflora and alien species, the spores of which happen to have been washed into the soil. Finally the method, while telling something of the species represented, tells us very little about their relative activity in the soil.

There are ingenious methods for infiltrating a litter/soil core with gelatin so that thin sections can be cut for examination under the microscope. Direct counts of bacterial numbers can be made and lengths of fungal hyphae can be measured. There are two major snags in this approach. You seldom know what species you are looking at and you cannot be sure whether the material you see is live or dead. Some bacterial counts made in this way have been several thousand times greater than those made with dilution plates. Are there huge numbers of dead bacteria in the soil? Or is the dilution plate count a gross underestimate? Burges (1958) makes two points in this connection. Firstly he draws our attention to the effects of protozoan predation which may limit the average life span of a bacterium to several hours. Secondly he notes that where dilution plate and direct count assessments have been made over the seasonal cycle the pattern of peaks and troughs is the same. Both point to the deficiencies of the dilution plate method. Nevertheless the method will continue to be widely used; for identifications must be made.

Recently a method has been developed for fungi making use of phase contrast in microscope inspection. Hyphal strands with protoplasmic contents can be picked out and these will provide the first direct estimates of active fungal biomass per unit volume.

Fungi active in a piece of wood or other large item of litter can be identified by first surface sterilizing it then placing a bit of it on suitable culture media. Fructifications observed in woodland may indicate which of such fungi are common. The next step is to study the decomposition rates achieved by these species under conditions similar to those in a wood.

There is still a long way to go; but with the stimulus of the I.B.P. considerable advances have been made on Macfadyen's (1963) expedient of subtracting the sum of all animal activity from estimates of total decomposer activity to arrive at a figure for saprophyte activity.

B Saprovores

For convenience and in accordance with past practice, this title is used loosely to cover all animals in the decomposer web that are not known to be predators or parasites. We have made the point that many of them are herbivores feeding on bacteria, spores and hyphae while others which consume decaying D.O.M. may be obtaining their nutriment from the same sources. This is not easy to check when minute animals are concerned for their faecal material does not yield identifiable bits of undigested organic matter. Cultivating a saprovore requires the discovery of an acceptable food but the latter is not necessarily the main item in its normal diet.

To date our knowledge of decomposer web animals is qualitative rather than quantitative and even the qualitative picture is rather sketchy. We are gradually amassing data about species that appear to be important in particular ecosystems. Micro-techniques in gas analysis allow us to make indirect estimates of the contribution they are making to decomposition. Thus Phillipson (1966) was able to measure the oxygen consumption of a single harvest spider (*Leiobunum rotundum*) and demonstrate the changing pattern of activity through the life cycle. Oxygen consumption is an index of activity and a crude indicator of the amount of food consumed.

Lebrun (1965) took monthly litter samples in a Belgian oakwood and estimated the mean abundance through the year of 46 species of Oribatid mites. In the laboratory he managed to culture 32 of these species, including all the more abundant ones; he obtained estimates of mean body weight and yearly oxygen consumption for each of them.

The data in Table 7.1 overleaf show that two species account for 60 per cent of the total activity. There are three major size groupings, suggesting three levels in the comminution spiral. Within each group very few species account for most of the activity. Note how activity per unit biomass decreases as the size of the individual increases—small 2.6, medium 1.2, large 0.6. This trend is a general one throughout the animal kingdom but there is considerable variation between taxonomic groups in the same size range. Lebrun therefore elected to interpolate the oxygen consumption of the species he was unable to culture using data from related species. In this way he arrived at his estimate of overall activity.

The estimation of saprovore population numbers requires many varied techniques partly because of the range in size and partly because of differences in mobility and behaviour. Scavengers on the surface, such as Carabid beetles and millipedes, roam over considerable areas: they can be caught in pit-fall traps and population numbers estimated by capture/recapture methods (see Chapter 8).

TABLE 7.1

The relative importance of Oribatid mite species in the decomposer web of an oakwood (from: Lebrun, 1965).

Importance Ranking	Small species 1–6·6 μg.		Medium spp. 15–90 μg.		Large spp. 156–620 μg.		Overall Ranking		
	no.	activity*	no.	activity	no.	activity	no.	activity	%
1	1	33	1	501	1	378	2	879	60
2	5	87	5	259	2	122	13	501	34
3	10	9	16	53	5	25	31	87	6
Totals	16	129	22	793	8	525	46	1 467	100
Biomass mg.		50		673		845		1 568	
Activity/unit biomass		2·6		1·2		0·6			

* Cubic millimetres of oxygen consumed per year is the index of physiological activity of the species population used here.

Segmented worms (earthworms and Enchytreid worms) can be persuaded to emerge from the soil by saturating it with dilute formalin or potassium permanganate. Micro-arthropods live in the spaces within litter or soil. They normally enjoy a humid micro-environment and, when the upper litter layer dries out during dry spells, they move to lower levels. This behaviour pattern can be exploited in an apparatus such as the Tullgren funnel, see Fig. 7.3.

A set of such funnels will be required and Macfadyen (1963) has shown that their efficiency is greatly improved if the lower parts are cooled. This becomes practicable if they are built as a battery of funnels and the part below the gauze is enclosed and air conditioned.

Other members of the mesofauna live in the waterfilm that normally covers soil particles. These require a method of wet extraction in which flotation is often useful to separate mineral particles from organic matter and organisms with unwettable exoskeletons (e.g. the micro-arthropods isolated in the Tullgren funnels) from water-film dwellers. The most important taxonomic group may be the nematodes.

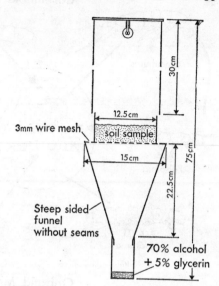

3mm wire mesh

soil sample

12.5cm

15cm

30 cm

75 cm

22.5 cm

Steep sided
funnel
without seams

70% alcohol
+ 5% glycerin

Fig. 7.3. Diagrammatic
section of a Tullgren funnel
(from: Jackson and Raw,
1966).

The microfauna comprise mainly Protozoa. Many species can be
isolated and identified using a modification of the dilution plate
method in which bacteria known to be palatable to protozoa are
incorporated in the agar medium. Testate amoebae can be estimated
by a direct count of their characteristic 'shells' which remain after
the live parts have been 'digested'. Wallwork (1970) discusses the
ecological work that has been done on protozoa (and other soil
animals).

This short account of the decomposers is by no means compre-
hensive. Hudson (1972) provides an excellent short account of the
fungi active in litter, wood and faecal material. The specialist
beetles that live in bark and wood have yet to be studied in an
ecosystem context, though much is known about those species which
have become pests (Forestry Commission publications).

Practical work—decomposer populations

Many of the techniques outlined above require expertise and
equipment only available at university level. However, Jackson and
Raw (1966) give clear diagrams of several pieces of apparatus that
could be constructed quite inexpensively for use in schools. Even a
home-made Tullgren funnel, for example, will yield an interesting

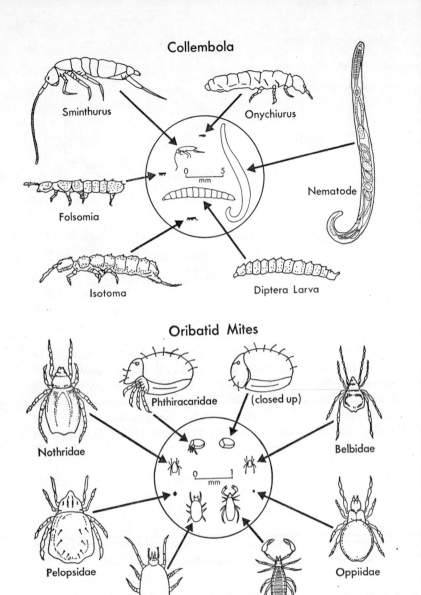

Fig. 7.4. Some common microorganisms of litter and soil.

Note that dimensions are indicated above merely to give some idea of relative size. There will be species and individuals in each of the groups included which are larger or smaller than indicated here.

assortment of springtails (*Collembola*) and mites (*Acari*) though extraction may not be efficient. Some common microorganisms of litter and soil are given in Fig. 7.4. If a battery of small Tullgren funnels can be made try one or both of the following investigations.

1. How do springtails adapt to their micro-environment?

Cut several cores of litter/soil in a coniferous wood with a well developed litter layer, divide each core horizontally into three slices and place each in a polythene bag. In the laboratory transfer each slice into a Tullgren funnel. A great variety of micro-arthropods will be extracted. The *Collembola* are larger than the mites and a good group, therefore, to study initially. Note the transition from small, white, blind species at the bottom to larger dark coloured species (with well-developed eyes) near the surface.

2. How are micro-arthropods dispersed?

Cut a large square section of litter/soil. In the laboratory divide it into as many squares as you have Tullgren funnels and extract the micro-arthropods. Sort out the springtails from the mites and then look for mites with conspicuous features. You will not be able to identify species as a rule. The mesostigmatid mites will be recognisable from their long legs—they are mobile predators, mainly of *Collembola*. Compare the numbers of each group identified in each sub-division of the large core. What can be said about the dispersion of species on a horizontal plane?

Using the formula for a Poisson distribution you could calculate the expected frequency distribution for the data in Fig. 7.5 overleaf if dispersion were random, though the calculation becomes excessively tedious when the mean ($\bar{x}$) is as high as it is for Nanorchestes. It is simpler to calculate the Index of Clustering (I) which has a value of zero when dispersion is random (the variance equals the mean in a Poisson distribution). Aggregation is pronounced for Onychiurus and even more so for Nanorchestes. The predatory mites, however, show a degree of clustering not significantly different from a random pattern (it will occur in three out of ten randomised dispersions of 74 mites among 42 sample units). Be careful, however, in interpreting data of this kind. The apparent type of dispersion may well change if you alter the size of the recording unit, (2 cm × 2 cm above). Indeed one method of determining the scale of pattern shown by populations is to calculate the 'variance' for the same data grouped

Fig. 7.5. Dispersion patterns of some easily recognised categories of microarthropod (collected in sets of 16×4 cm² sample units from pinewood litter).

ONYCHIURUS (Collembola—Springtails)—probably one species only

0	18	3	0	0	2	0	0	14	11	0	0	0	1	2	1	0	0
0	9	2	2	11	0	11	1	5	27	3	0	2	2	0	0	0	2
2	0	4	1	5	0	0	0	1	1	0	0	2	0	0	1	0	7
7	0	1	0	16	35	1	6	0	1	0	2	1	6	2	1	1	0

n 80
x̄ 3·075
V 35·8
I 10·7

MESOSTIGMATA (Acari—Mites)—several genera of this predator group*

—	1	—	1	3	—	8	—	—	—	0	—	3
3	3	0	4	—	0	4	3	—	1	—	0	—
0	0	1	0	2	—	3	1	—	0	1	1	2
4	2	3	—	0	2	—	0	—	2	—	2	1

n 42
x̄ 1·762
V 2·576
I 0·5

NANORCHESTES arboriger (Acari)—a distinctively coloured herbivorous species

7	28	4	8	6	35	12	6	11	17	78	19	22	35	17	8	11	23	8	0
13	24	37	4	11	33	8	27	0	0	0	31	24	20	21	8	2	53	0	22
21	6	2	8	0	9	13	89	76	17	47	24	42	10	4	17	12	7	0	12
7	4	11	15	20	37	8	32	25	13	8	0	19	0	120	91	17	24	35	7

n 80
x̄ 20·025
V 496·2
I 23·8

n = no. of sample units
x̄ = mean no. per sample unit

x = no. of organisms in a sample unit
$$V = \text{variance} = \frac{\Sigma\,(x-\bar{x})^2}{n-1}$$

$$\text{Index of Clustering } (I) = \frac{V}{\bar{x}} - 1$$

*analysis incomplete

into successively larger units. Variance rises markedly above the trend line when the unit size approximates to the predominant population pattern unit size. This approach has been used extensively in the study of vegetation patterns (Kershaw 1964).

Consider the possible ecological implications of both random dispersion and clustering at various scales.

3. Estimating the biomass of worms.

You are only likely to find worms in any numbers in woodland with brown forest soils. Choose an ash, sycamore or oakwood (if the latter, one with a reasonably rich groundflora). Clear away the litter and surface twigs on several patches and mark out squares with half-metre sides. Prepare a dilute solution of formalin (25 ccs of 40% formalin to 4·5 litres of water or 2 per cent potassium permanganate in a watering can and spray the quadrat until the soil is saturated. Collect the earthworms as they appear over a period of 15–20 minutes. Let them come completely out before you pick them up or they will slip from your grasp and retreat into their holes. Preserve them in formalin. In the laboratory wash them in water, dry with blotting paper and weigh. Assume that some ten per cent of the fresh weight will have been lost while in formalin. Now calculate the fresh weight of worms per unit area. If you have done a similar exercise in pasture or on a lawn, there will be interesting comparisons to make between the species that are most abundant. As with many animals, immature worms are difficult to identify— these are the ones without the swollen segments that constitute the *clitellum*. For this reason the project is best carried out in summer or early autumn. You may also find that Enchytraeid worms come to the surface: they are much smaller, seldom exceeding a centimetre or so in length—a lens is required to see that they are indeed segmented worms.

4. Following the decay of branches or stumps.

To do this observations are required over a period of a year or more. Visits should be timed to follow damp spells when wood and bark organisms will have been active and may be producing fructifications.

On fallen beech branches, for example, you may find that a succesion of different fungi takes place. The coral spot fungus (*Nectria cinnabarina*) may appear first, followed by *Hypoxylon fragiforme* or *Diatrype disciformis* and then *Stereum hirsutum* with its 'shelves' of yellowish bracket-like fructifications, or the not dissimilar *Trametes versicolor* which has multicoloured fructifications with pores on the undersurface.

Mark a number of branches and stumps for systematic observation along a woodland path. Compare the sequence of fungi on the branches of different species and on branches of different size. Investigate the kind of decomposition that takes place. Some species are lignin decomposers which leave white punky 'wood' while others which take mainly cellulose cause brown rots. *Xylaria hypoxylon* which has antler-like fructifications causes a white rot bordered by black '*zone-lines*' where its hyphae have come into competition with another mycelium.

8 Assessing the relative importance of species populations—herbivores, predators and parasites

III Herbivores

Herbivores are found where the most palatable food is available—in the tree canopy, the ground vegetation and the top soil horizon where most of the fine roots occur. Each location has its characteristic species.

A The tree canopy

Most woodland tree species are liable to have their foliage partially devoured by herbivorous insects, mainly caterpillars (larvae of Lepidoptera). Bray (1964) measured the loss of photosynthetic area in a sample of leaves from three broad-leaved woods in America and concluded that it was of the order of eight per cent. More recent work suggests that the leaf area lost is usually much less than this. Periodically much heavier losses occur. Oakwoods for example, suffer virtual defoliation by caterpillars of moths (*Tortrix* spp.) at irregular intervals. Primary production is drastically reduced, biomass increase is cut right back and full fruiting is delayed, perhaps for several years. The effect is mitigated slightly by the ability of oak to produce a second (*lammas*) crop of leaves. Pine plantations have occasionally failed to recover from defoliation by the 'looper' caterpillar of the moth, *Bupalus piniarius*.

A knowledge of the biology of each species of defoliator is essential if the most efficient means of monitoring its numbers is to be devised. For *Bupalus*, for example, a count can be made each winter of the number of pupae in the pinewood litter. Experience suggests that numbers of five pupae per square metre or less indicate a looper population at, or below, *endemic level*. Higher numbers are danger signals for a build-up to *epiphytotic* levels ('epidemic', not on people, but on plants), though as many as 20 per square metre have been recorded in a winter not followed by severe defoliation next summer (see also Chapter 9). Rather naturally a great deal more research has been done on species capable of reaching epiphytotic proportions on plants of economic importance to man, even though, at other times, they may make an unimportant contribution to ecosystem metabolism.

Foliage losses occur in other ways. Twigs may be weakened by shoot-boring larvae (e.g. *Myelophilus* beetle larvae on pine): the whole twig with its leaves may then snap off in high winds. Leaf-miners (larvae of insects in several major orders) eat out the meso-phyll tissue of the leaf. Black game (*Lyrurus tetrix*) feed on the buds and young shoots of conifers.

The fruit and seeds of trees are the main food source of another group of species of which the squirrels (*Sciurus* spp.) are possibly the most important.

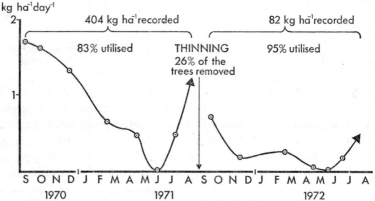

Fig. 8.1. Pattern of cone fall as recorded in litter collections. The weights recorded are of cone dry weight either as whole cones or as cone scales and cores. The seeds are assumed to have been eaten when the cone is broken into pieces.

Figure 8.1 records the seasonal change in dry weight of cone material from which red squirrels (*S. vulgaris*) have extracted the seed. It shows clearly the 'low' in late spring and early summer when squirrels are forced to look for other kinds of food. This is when they take birds' eggs, fledglings, tree buds and live bark. In the forester's eyes the bark damage, often on the leading shoot, is serious and squirrels are regarded as pests in plantations. The two sets of 12 month figures above are not comparable because of an intervening thinning operation by foresters and a decline in cone production. But the data show clearly that the intensity of cone utilisation

increases when the amount of cones available falls. There are general
implications for seed crops here: the heavier the crop the smaller
the percentage taken by herbivores. This is one reason why tree
seedlings appear in large numbers only intermittently (the 'regeneration waves' referred to in Chapter 3). Squirrels feed partly on fallen
cones and may be trapped on the ground. Table 8.1 overleaf shows
the results that were obtained using 40 randomly located traps in a
wood 130 hectares in extent.

Each trapping period began with five to seven days when the traps
were baited with maize but not set. Then followed a week of intensive
trapping with visits twice daily. All occupants were recorded, marked
if captured for the first time, and then released. You will notice that
the population estimates are rather crude—the most reliable estimate
(March 1967) lying between 92 and 124. Over half the population had
been marked before this precision was reached. Nevertheless it is
clear that there are seasonal fluctuations. Numbers rise with the
appearance of the new broods in autumn but winter mortality is
heavy, and an unknown number emigrate.

B The ground vegetation

In managed conifer woodland dense herbaceous vegetation will be
confined to rides, roadside verges and new plantations in their first
few years before the canopy closes. Deciduous woods in their mature
phase may be sufficiently open to support a grassy groundflora
(soft-grass, *Holcus mollis*, and bent, *Agrostis canina*, are the likely
dominants). These are the sorts of places in which roe deer (*Capreolus
capreolus*) and fallow deer (*Dama dama*) may be found grazing in
early morning or at dusk. They lie up during the day choosing the
protection that thickets of young tree growth afford. The extensive
re-afforestation of recent decades has provided an opportunity for
both of these attractive animals to establish themselves over most of
Britain in increasing numbers. Roe deer are territorial animals which
means that the breeding population in any wood fluctuates within
pretty well defined limits set by the food resources available. They
are true woodland animals and seldom venture out into agricultural
land. Not so the fallow deer which have caused extensive damage to
crops when their numbers were not carefully controlled. Deer
management is a specialist job, for the assessment of population size
requires the ability to recognise all the breeding individuals. Females

TABLE 8.1

Capture/recapture data—red squirrel (*Sciurus vulgaris*) Edensmuir Wood, Fife, Scotland.

Trapping Sessions	May '65	Apr. '66	July '66	Nov. '66	Jan. '67	Mar. '67	May '67	July '67	Sept. '67	Nov. '67	Jan. '68	Mar. '68	May '68
New markings followed by the number of subsequent recaptures.	20	4	5	4	6	5	4	4	1	0	0	3	3
		18	3	2	3	4	5	4	0	0	1	1	2
			26	8	14	15	11	10	1	2	4	7	6
				6	2	3	2	2	3	1	1	3	0
					10	8	8	3	2	0	2	1	2
						11	8	4	2	0	1	2	3
							6	1	0	0	0	0	1
							8	3	0	1	0	1	1
								10	1	1	2	4	4
									11	**5**	1	2	3
											2	2	2
												0	**7**
												11	(17)

Pop. Estimate (N) = a × b/r

Variance Estimate (V) = N × $\dfrac{a(b-r)}{r^2}$

Confidence Limit % = $\sqrt{V} \times t \times \dfrac{100}{N}$ (t is Student's 't')

	May '65	Apr. '66	July '66	Nov. '66	Jan. '67	Mar. '67	May '67	July '67	Sept. '67	Nov. '67	Jan. '68	Mar. '68	May '68
Prior total marked* (a)	0	20	36	62	68	77	88	94	102	113	117	118	129
Trapped in session (b)	20	22	34	20	35	46	44	39	18	10	14	37	51
Recaptures (r)	—	4	8	14	25	35	36	29	7	5	12	26	34
Population estimate		110	153	89	95	101	108	126	262	226	137	168	194
Confidence limits %	±	85	62	29	22	17	15	19	59	63	24	21	20

* Less known mortality
(from A. Tittensor, 1970)

are culled to control numbers and males to control quality and limit damage. Red deer (*Cervus elaphus*) are forest dwellers in Europe. In Britain this is just beginning to happen again after centuries of living mainly above the tree line and coming down into woods for shelter and browse during the more severe winter months. Visitations of this kind (c.f. migrant bird visitors) are a problem to the ecosystem analyst: it is very difficult to assess their contribution to ecosystem metabolism.

Feeding in much the same areas of woods are several small rodents of which the commonest are likely to be the wood mouse (*Apodemus sylvaticus*) or the bank vole (*Clethrionomys glareolus*). Their populations have also been assessed by capture/recapture methods. The traps used are smaller than those used for squirrels and the bait is mashed oatmeal. Southern (1959) followed population fluctuations of wood mice and bank voles for ten years in a 100 hectare area of Wytham Wood near Oxford. During this period it was discovered that the social organisation in the vole population was invalidating some of the assumptions on which the capture/recapture method is based. There is a social hierarchy and individuals lower in the hierarchy are more frequent visitors to traps. Total numbers are likely to have been considerably underestimated. 'Trap addiction' by certain individuals can be spotted when the same animal turns up many times in one trapping session (only scored as one recapture). But there is no way of judging the extent of 'trap avoidance' as a behavioural characteristic unless there are alternative data to provide population estimates.

C Fine roots

Here ecology is still in the natural history phase, the collection of observations which indicate that roots may be an important energy source for particular species.

IV Predators

We have seen that the population numbers of herbivores may sometimes increase to epiphytotic levels, over-exploiting their food source and reducing its potential to support them. The elimination of their predators is one way to ensure that this will happen with greater frequency. When predator/prey relationships were first considered

D

it seemed obvious that predators controlled prey numbers. Now we are not so sure. It is clear that predators breed more prolifically if the numbers of their prey increase, leading to an increase in predation the next year—a *delayed numerical response*. Equally if prey numbers decline there will be a delayed response in predator numbers. But it is difficult to find evidence that predators can cause a significant decline in prey numbers once the prey species has become super-abundant. The few well-known examples provide spectacular examples of biological control. An Australian bug, the cottony cushion scale insect (*Icerya purchasi*) had been inadvertently introduced into the Californian citrus orchards where it became a serious pest. In December 1888, 129 adults of its Australian ladybird predator (*Novius cardinalis*) were released within a cage over a heavily infested tree. All the scale insects were killed by April 1889. The cage was opened and the orchard was free of *Icerya* by July. Within a few years the bug was no longer a pest anywhere in California. Note that both species are still present in California but at endemic levels: if the bug were exterminated then the ladybird would also die out unless it found alternative prey. Unfortunately this kind of predator/prey relationship is the exception rather than the rule. It is mostly found among invertebrates.

Woodlands normally have *mixed predation systems* in which several predator species will take, at various times, any of several alternative prey species. This system allows all the predator species to turn their attention to whichever prey species is particularly abundant at the time. Food gathering is thus easier for the predator, the numbers of the most abundant prey are reduced and the prey species currently at very low population levels have a respite from predation. The predators show a combined, immediate *functional response* to an increase in the numbers of one of their prey species. In terms of behaviour they are said to form a *searching image* for prey that is abundant and at the same time predation on other prey species falls below the *random encounter level*. There can be little doubt that mixed predation systems, in conjunction with other environmental factors causing mortality, do damp down fluctuations in prey numbers. At the usual endemic levels this will amount to control! When a prey species achieves superabundance Man has often been responsible directly or indirectly for creating conditions so favourable for the rapid increase in numbers that ecosystem control mechanisms became inadequate. There are exceptions to this statement. In boreal regions the numbers of many herbivores seem to oscillate with considerable regularity between superabundance and comparative rarity without Man's intervention. The fact that this phenomenon occurs is sometimes attributed to the lack of diversity in ecosystems of this region; though it would seem that the argument is a circular one, for it is based on the assumption that the less diverse an ecosystem the fewer the *homeostatic* (control) mechanisms operating. Certainly the down-turn in number of herbivores, when it occurs, is not caused by predation. The snow-shoe rabbit or varying hare (*Lepus americanus*) has been described at the height of its cycle as being so abundant that you could hardly avoid treading on one: then the next year there is not one to be seen. The cause seems to be a combination of stress and disease. The story is similar for short-tailed voles (*Microtus agrestis*) in young plantations in Britain though the oscillations are not so marked. The exact role of predation has not been worked out. It is interesting to note that in a mixed predation system it is the concerted action of *all* the predator species that is the vital contribution to ecosystem homeostasis.

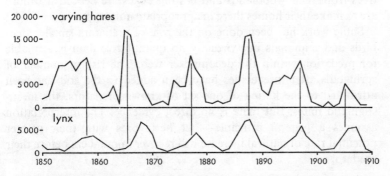

Fig. 8.2. The relative abundance of *Lepus americanus* and one of its predators over a sixty year period in Canada as indicated by trapping returns of the Hudson Bay Company (from: Hewitt in Elton, 1966).

British woodlands lost their last large mammalian predator, the wolf (*Canis lupus*) several centuries ago. Polecats (*Mustela putorius*) pine martens (*Martes martes*) and wild cats (*Felis silvestris*) linger on. Birds of prey (raptors) began to be systematically killed when game bird shooting became popular. More recently some raptorial birds have declined in numbers through the food-chain concentration of organo-chlorine compounds such as dieldrin and D.D.T. Incidentally birds provide the supreme example of what can be done in the way of population monitoring with the help of amateur naturalists. Reasonable estimates of the numbers of very many of our bird species are available only because of the continuing surveillance organised by Ornithological Societies all over Britain. Special research projects have been concerned with rare raptors such as the osprey (*Pandion haliaetus*) and the golden eagle (*Aquila chrysaetos*). However there have been few studies of raptorial birds in an ecosystem context. Perhaps the best known is that of Southern on the tawny owl (*Stryx alugo*) in Wytham Wood. In addition to the monitoring of population changes, Southern analysed the faecal pellets of the owl in which bones of mice and voles could be recognised. The predation rate of owls on small rodents could then be related to prey numbers, though at the time Southern was more interested in the way the interaction affected the breeding behaviour of the owl. Tinbergen's (1960) classic studies of titmice (*Parus* spp.) in Dutch woods has led to an appreciation of the role these small birds can play in reducing the numbers of woodland insects. Foresters tend to eliminate hollow trees from their woodlands and nesting boxes are needed if titmice are to make their homes there in appropriate numbers.

Some work has been done on the role of predators smaller than birds and mammals but virtually no quantitative data is available for predation within the decomposer web. The rise in number of springtails after insecticides have been added to the soil has been attributed to the killing of one set of known predators, the mesostigmatid mites. But there is no direct evidence. The interpretation assumes a general principle—that herbivores with their greater potential rate of natural increase will recover in numbers before their predators.

V Parasites

For convenience the remaining biota of the ecosystem will be dealt with under this heading. The term 'parasite' is, in any case, very loosely used including a variety of very different trophic relationships. There are exceptions to almost every generalization you can make:

1. Parasites are smaller than their hosts—but those such as the bird's nest orchid (*Neottia nidus-avis*) use soil/litter fungi as their energy source (before the fungal relationship was discovered they were described as saprophytes).

2. Parasites are closely associated with their hosts and are highly adapted to an environment inside the host or on its surface—but some blood sucking organisms (e.g. mosquitoes, leeches, vampire bats) are free living.

3. Parasites' adaptations tend to make them host specific, attacking only one species or several species in a fairly close taxonomic grouping—exceptions among some free-living parasites.

4. Parasites do not usually kill their host or even prevent it reproducing—exceptions are those parasitoids which lay eggs in the host tissues and eventually kill it e.g. Ichneumon flies.

5. Parasites' energy source derives from the living tissues of the host—but there are

(a) ectoparasites such as bird lice (*Mallophaga*) which are really scavengers feeding on sloughed dead organic matter, and

(b) endoparasites in those body cavities with openings to the outside, which are more properly classed as commensals, taking some of the food that would otherwise be available to the host.

Most of these generalizations also apply to one of the partners in mutualistic relationships in which advantages accrue to both partners. If the term **mutualism** is used, then the term **symbiosis** can be used in its original meaning of two species in intimate association. Symbiosis is then a feature of most parasitic as well as mutualistic relationships. A study of mutualistic relations shows clearly that many have evolved from essentially parasitic relationships and that some amount to reciprocal parasitism separated in time (e.g. the nodule forming, nitrogen fixing *Rhizobium* and its leguminous host). The most important examples of mutualism in woodland ecosystems are provided by mycorrhizae. Most trees are mycorrhizal to the extent that only a minority of fine roots have root hairs. In a wide range of soils of intermediate fertility it can readily be demonstrated that young trees with mycorrhizae put on dry weight faster than those without. Exactly how the tree benefits is still largely a matter for conjecture. In Chapter 5 it was suggested that mycorrhizae were likely to be more efficient than root hairs at uptake of nutrient ions in dilute solution. There is also the possibility that mycorrhizae effectively short-circuit the nutrient cycle through the fungal symbionts' ability to utilize nitrogen and phosphorus in organic form.

Are they responsible for nitrogen uptake in ammonia form? Certainly there are many woodland mor-humus soils in which conditions are quite unsuitable for nitrifying bacteria, making it unlikely that nitrate is the main form of nitrogen uptake by trees there.

Parasites, then, are normally symbiotic with their hosts and are identified and assessed by examination of the host. Their importance in ecosystem metabolism, which is our main interest here, depends on the effect their activity has on the host population and this is another reason for starting with the host rather than with free living stages of the parasite.

The approach we have used so far, through population numbers, biomass and average activity is not readily applicable to many parasites. Some of the larger ectoparasites such as body lice (*Pediculus humanus*) can be treated in this way. However the majority of parasites and particularly endoparasitic micro-organisms cannot be cultured in abiotic media to determine metabolic rates. Their activity has to be assessed from the host/parasite interaction, from the symptoms of attack. If there are no symptoms the presence of a parasite will not be suspected. In theory it should be possible to determine the reduction in host productivity caused by a parasite by comparing the performance of parasitised and non-parasitised populations of the host. Indeed this has been done with many 'economic' species in managed resource systems where the crop or livestock are substantially free from pests and disease most of the time. In other situations an individual is likely to be attacked with varying severity by several parasites at the same time. Debility or even mortality can seldom be attributed with confidence to one parasite alone. Physical environment factors may cause mortality or render the host more susceptible to parasite attack. Thus the effect of a particular parasite on the host is often a matter of opinion. For example, we find Andrewartha and Birch (1954) attributing the mortality of the grasshopper, *Austroicetes cruciata*, in the southern part of its range in South Australia to wet weather, although the dead grasshoppers had clearly been attacked by a parasitic fungus. The effects of parasitism on animals are also linked with predation, for heavily parasitised individuals are more likely to be taken as prey. An opposite effect may obtain with herbivores, if healthy plants are preferred to diseased plants as food.

The above remarks do not apply to parasitoids, which kill their hosts. The egg develops symbiotically in host tissue and eventually after the host's death the parasitoid emerges leaving a characteristic exit hole. It is often possible to attribute host mortality to a particular species of parasitoid. An example will be discussed in the next chapter.

Practical work—herbivores

A person setting out to estimate the population numbers of a woodland animal would begin with long periods of quiet observation—learning to recognise them in all their life phases, to differentiate between the sexes—learning where and when they feed and the nature of that food. Perhaps he would need to work at night or creep in place before dawn. Certainly the work does not fit happily into class work periods, nor is the necessary expertise likely to be attainable in class time. A group of people invading a wood will be lucky to get more than a fleeting glimpse of some of the animals and birds that are there. Generally the kind of work a class will be advised to attempt is the gathering of indirect evidence of animal activity, the clues they leave behind. Look for tracks where the ground is soft, for evidence of grazing and browsing, for fruits and seeds which have been partly eaten, for droppings and for nests. Squirrel dreys can be counted, though good organisation is necessary if the count is to be accurate—dreys can be distinguished from the nests of birds by their spherical rather than flattened shape. The class moves through a defined part of the wood in line, each person checking with his neighbour whenever he sites a drey to make sure it is not counted twice. One family of squirrels commonly has several dreys of which perhaps only two are in use: very roughly there will be as many squirrels as there are dreys.

If you are making litter collections regularly in a wood there may be evidence in them of animal activity. We have mentioned pine cones from which the seeds have been taken by squirrels or sometimes crossbills (*Loxia* spp.). Live shoots that have fallen because of shoot-borer activity can easily be recognised and noted as an index of shoot-borer numbers from year to year.

The trapping of small rodents might be demonstrated. As a matter of principle students should not handle the animals or mark them except under expert supervision. All that is suggested is the setting up of a line of traps* 10 to 15 metres apart. Skill is needed in recognising places where small rodents are likely to feed, in placing the trap on one of their runs and camouflaging it effectively. Baiting is not required but some mashed oatmeal should be placed in the trap itself; for these small animals need to be feeding more or less continuously to stay alive. The traps must be visited twice a day and obviously they must be very close to hand if trap rounds have to be fitted in with other work.

If you have a pine wood nearby the litter might be sampled in winter for pupae present. A random layout of quadrats is suggested, the litter being searched on the spot and replaced as it was. Forestry Commission leaflet no. 40 has colour pictures of the various kinds of pupae you may find. Is the number of *Bupalus* pupae dangerously high?

* Longworth small mammal traps currently cost about £3 each.

9 Predictive models

Predictions about a system can be arrived at in two ways—by precedent at the operative level or by functional integration of systems belonging to lower levels. The first demands no knowledge of how the system works: it is the *black-box* approach of the physicist in which outputs are related to variation in inputs. Mathematically the relationships are expressed simply in multiple regression equations.

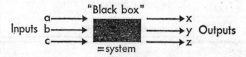

Fig. 9.1. The black-box diagram.

The second way demands reasonable knowledge of how the system works, how its processes interact, the synthesis approach. Processes at one level are of course systems in their own right at the level below.

In the diagram overleaf the ecosystem is shown as composed of three interacting sub-systems or processes, primary production, consumption of live organic matter and consumption of dead organic matter. Process rates are now seen to be dependent not only on the inputs, the *driving variables*, but also *state variables*, the levels in various biomass and dead organic stores in the system. Thus primary production achieved depends on inputs of radiation and carbon dioxide but it also depends on the biomass of green plants. Note the return arrows representing feed-back mechanisms: herbivore activity, for example, reduces the state variable, biomass, and so the rate of primary production. Each of these relationships must be expressed mathematically. The amount of detail will vary with the state of our knowledge. The herbivore web activity could perhaps be represented by population system data for the dominant herbivores and predators. We have already noted that such an approach is not

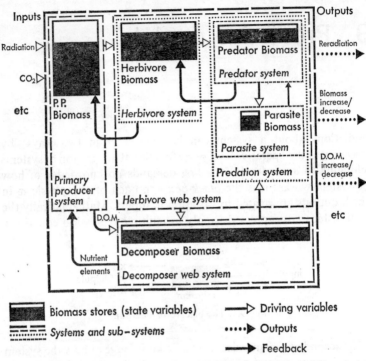

Fig. 9.2. Interrelationships between subsystems of the ecosystem.

yet feasible for decomposition (Chapter 7). Primary production has been studied intensively as a system in its own right at several levels, the leaf, the plant and the population (or crop). An early example at the population level was a study by De Wit (1963) at Wageningen in Holland:

"The photosynthesis rate of a leaf canopy depends on the reflection, the transmission and the photosynthesis function of leaves, the position of the leaves with respect to the horizontal surface and each other, the leaf area per unit soil area, the amount of diffuse and direct light, the height of the sun and the resistance against the transfer of carbon dioxide from the bulk of the air to the canopy. . . . The solution has been carried to the stage where the daily photosynthesis of a canopy with known characteristics can be computed for any time and place on earth from the relevant meteorological data."

Of course De Wit's ecosystem was a fertilised arable crop, weeded and sprayed as necessary to limit competition and herbivore or parasite activity. But it will not be long before we know enough about tree canopies to apply the same 'systems analysis' approach to the tree component of woodland ecosystems. In the computer we have a tool well able to cope with mathematical relationships of this degree of complexity.

If such a primary production system model were available for a woodland ecosystem it could be incorporated in the ecosystem model. However the first *simulation model* is usually kept as simple as possible. Its computer programme is run with various combinations of inputs and the predicted outputs compared with observation. If the model behaves reasonably like the system it is said to be validated. Now it can be analysed for the sensitivity of its component processes (equations): this is exactly what it sounds like, the identification of the processes where a small change in an input parameter has a comparatively large effect on an ecosystem output parameter. These are the ones where further research and consequent refinement of the model would be most rewarding. The model can be used to predict out-turns, should so far unobserved combinations of driving variable occur. It can be run repetitively to investigate the long term effect of small changes, though this kind of exercise must be interpreted very carefully: perhaps most will be learnt from the nonsense answers it produces.

Studies of this kind at the ecosystem level are as yet in their infancy and there is much to be learnt from systems analysis and modelling at other levels and in other disciplines. The name **Systems Ecology** has been given to this field of study.

Prediction by precedent will remain our main technique for some time. Let us look briefly at what has been done to date and consider some of the implications. The Forestry Commission have compiled the results of some 40 years research in their Management Tables (F.C. Booklet No. 34): these specify the quantity and range of size of produce available at any time in the crop development of the main plantation species used in Britain. A rather remarkable generalisation emerges from this work: the development pattern on fertile and infertile sites is the same, only the rate of development is different. One curve will represent, for example, the biomass increase of both slow and fast grown crops if biomass is plotted against a height parameter rather than time. Now the Commission data are rather specialised, giving volume production (not dry weight) for stems only (no branch or root estimates). A woodland ecologist who concludes that the data are therefore useless to him is throwing away a major advantage he has over ecologists working in other systems. All he needs to do to make those data extremely useful is:

1. Record some of the tree crop parameters in the same way as the Commission. At its simplest this entails estimating *top height* (see Chapter 6) which in turn requires the identification of the largest girthed trees. In this context it is interesting to note the importance attached to standardisation of measurements and techniques in the recent International Biological Programme, for only then can valid comparisons be readily made (Newbould 1967, IBP Handbook No. 2).

2. Estimate the mean specific gravity of tree stems in the wood. This is most simply done in conjunction with the estimation of the dry weight of sample trees (Chapter 6): each stem section, after weighing, is measured for mid girth over bark and length; stem section volume is assumed to approximate to a cylinder. Note that you have determined 'green volume'. As soon as the stem sections begin to dry out they shrink radially and specific gravity determined from air-dry logs will be appreciably higher.

He can now compare his own estimates of stem production with yield table estimates, and because of the close correlation between many crop parameters the trends evident in the Commission data for stems are likely to apply to total biomass. Table 9.1 shows the data he might use for comparison if his pine plantation were 38 years old and had a top height of just under 12 metres.

The very fact that prediction by precedent in the form of yield tables has been so successful points to several features which most woodland ecosystems share. They are large, complex and remarkably stable ecosystems in which primary production accumulates at a steady rate, irrespective, almost, of the yearly fluctuations in the driving variables. This applies to plantations as well as woods of more natural origin, though, occasionally, the poor choice of species for a particular site will result in stresses that lead to instability in ecosystem metabolism and consequent failure.

Woodland populations of organisms in the herbivore and decomposer webs are by implication fluctuating about a steady level, their

TABLE 9.1

Extracts from Forestry Commission Yield Class Data for pine (Forestry Commission, 1971)

Yield Class	Main Crop after Thinning						Thinnings		Cumulative productivity		
	Age Yrs.	Top Ht. m.	No. per ha.	B.A. m²/ha.	Girth (MBA tree) cm	Volume m³/ha.	No. per ha.	Vol. m³/ha.	Stem Prod. m³/ha.	M.A.P. m³/ha./yr.	C.A.P. m³/ha./yr.
8	35	12·0	1549	19·9	40·2	116·3	645	33·5	199·0	5·69	11·28
	40	13·5	1189	22·0	48·1	141·3	360	31·4	255·4	6·39	11·50
	45	14·9	960	24·3	56·5	168·6	229	30·2	312·9	6·95	
6	35	9·8	2494	19·6	31·4	94·4	1336	30·2	135·2	3·86	8·32
	40	11·2	1834	19·8	36·8	109·9	660	26·1	176·8	4·42	8·78
	45	12·5	1446	21·3	43·0	129·4	388	24·4	220·7	4·90	

Notes:
1. These data refer only to stems and include bark. Table 53 in F. C. Booklet No. 34 has been used to interpolate volume to the tip of the stem.
2. M.A.P. is the mean annual production over the life of the crop.
3. C.A.P. is current annual production here estimated from the five year periodic mean annual production.
4. Volume in cubic metres multiplied by Specific Gravity gives dry weight in tonnes.

activity in toto seldom affecting the long term trends in primary productivity. The assumption is sometimes made that a population has been under stress if an outbreak of pests or disease above endemic levels occurs. But this ignores the possibility that predation may have fallen to exceptionally low levels: or that an exceptionally favourable weather sequence for a pest has recurred in several successive years. We need enough knowledge of the ecology of each potential pest species to identify the factors responsible. Why is it, for example, that epiphytotics of the pine looper moth have devastated whole stands in England and Germany, while looper populations have remained at endemic levels in Holland?

Klomp (1966) studied the looper population in 6·8 hectares of a pine wood in Holland over a 15 year period. He assessed the numbers of nine different stages in the life cycle each year and whenever possible identified the causes of mortality between them. The results are shown in what is called a *Life Table*.

TABLE 9.2

Extracts from the life table for *Bupalus piniarius* (from: Klomp, 1966).

Stage	Assessed	1955–56 Density m^{-2}	Mortality ascribed to	1962–63 Density m^{-2}	Mortality ascribed to
Eggs	July	33	9·3 Trichogramma (parasitoid)	99	15·7 Trichogramma
Larvae I	July	22·5		77	
Larvae II–III	Aug.	13·9		24·8	
Larvae III–IV	Sept.	12·0	0·1 Parasites	20·1	0·6 Parasites
Larvae IV–VI	Oct.	7·4		14·8	
Nymphs	Nov.	3·2		8·4	
Pupae	Dec.	2·7	0·08 Winter predation	5·5	0·5 Winter predation
Pupae	April	2·6	0·82 Parasites	5·0	2·0 Parasites
Moths	June	1·5	(55·3%♂)	1·1	(50·8%♂)
Females	June	0·67		0·54	
Females	July	0·35*		0·26*	
Eggs		58		47	

*This mortality among female moths is deduced by comparing egg laying potential with eggs actually laid.

The complete data are instructive, indicating the extent to which population numbers fluctuate about a steady level. These fluctuations for numbers of eggs and adults are shown in Fig. 9.3 below.

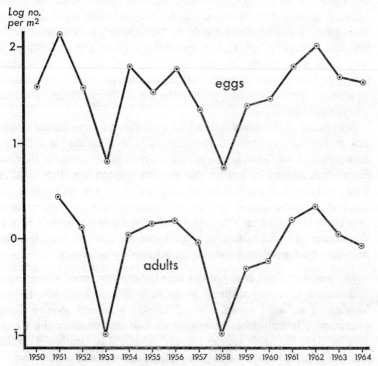

Fig. 9.3. Pine looper moth *Bupalus piniarius* L. Density variation over 14 years (from: Klomp, 1966).

Adults averaged about one per square metre with lows around one per ten square metres and highs around five per two square metres, a twenty-five-fold range. Larvae in September (at their most voracious stage) reached 25 per square metre on two occasions. However this only amounted to one September larva to each 18 first year pine shoots. Obviously food was in no way limiting and, equally, the trees did not lose too much of their photosynthetic apparatus through *Bupalus* activity. Now it is readily understandable that looper populations should recover after being reduced to low levels: predation is then likely to be below random encounter level (see Chapter 8) and each adult female can lay over two hundred eggs.

What is much more difficult to account for is the downturn, when population numbers are high: mortality factors must be increasing in intensity as population numbers rise. Mortality of this kind is distinguished as *density dependent*. Klomp found that mortality between generations operated in this way, though total mortality was always high: numbers would be static with a mortality of 99·54 per cent. When he investigated mortality between life-cycle stages it was clear that nearly all this mortality was *density independent* (unrelated to density changes) or merely *density related* (increasing in amount but not intensity). Only the mortality of late larval stages was density dependent.

It appears as if density dependent mortality of large larvae would not in itself be enough to regulate *Bupalus* numbers: it must be accompanied by very high mortality from other causes. Further research is needed before we can predict epiphytotic outbreaks. A study of large larvae predation should reveal the species responsible for, and the mechanics of, the density dependent mortality: were these predators absent or under-represented in the disaster areas? If not, the answer must lie in factors responsible for the very high level of non-density-dependent mortality in Klomp's pine wood.

No practical work is suggested here for the very good reason that it demands resources unlikely to be immediately available. If you happen to be working in a wood in which several species share dominance, it would be interesting to find out whether the Commission data for single species plantations could be combined on the basis of relative basal area to give realistic estimates of production.

10 Woodland types

Let us suppose you have done some woodland research and are about to discuss your results with someone who has worked in woodlands elsewhere. You will start by trying to tell him as succinctly as possible what kind of woodland you worked in. You must use terms (classification labels) that he will understand. This is the reason why scientists all over the world use binomial Latin names to describe the species they work with. Unfortunately the classification of vegetation is not so straightforward as the taxonomy of organisms. There has been much dispute as to the precise definition of a species but it is clear that there is a genetic basis for its existence as an entity. Amongst vegetation ecologists the dispute about the basic units for vegetation classification continues: some claim there is no logical reason why they should exist. There is not space here to deal with the many and varied developments in vegetational theory (but see Shimwell 1971). We will look briefly at the ideas behind currently popular methods of vegetation analysis.

Most ecologists would agree that similar assemblages of species turn up repeatedly on similar sites within a climatic region. They would also agree that no two assemblages were exactly the same. In places an assemblage is a visually discrete entity with abrupt ecotones between it and neighbouring assemblages. Elsewhere one kind of assemblage will grade insensibly into others. Some assemblages are floristically rich; others include very few species.

To make a systematic investigation of floristic composition you could lay down a transect and record each new species as you move along. Subsequently the cumulative number of species could be plotted against area covered to give a *species/area curve*. Most of the common species are encountered near the beginning and the curve rises steeply. The further along the traverse you go the fewer new species you find—the curve is flattening out. However, as soon as the transect moves into a different assemblage of species there will be a new upsurge in numbers.

Figure 10.1 provides an example of such a new upsurge in species number that occurred when a transect through an alder wood ran into a streamside community. Have we here a new woodland type? Or should we regard streamsides as part of the normal heterogeneity to be expected in an alder wood? There is no clear-cut answer to this dilemma. The way it is resolved depends on

1. the number of new species encountered

2. the number of already recorded species present

3. the area occupied by the streamside community

4. the approach to vegetation analysis that you have inherited.

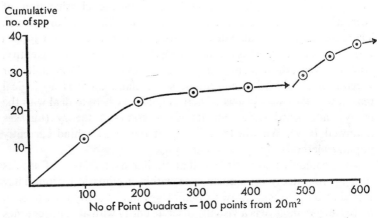

Fig. 10.1. A species/area curve compiled from point quadrat data along a transect in an alder wood. One five point frame per square metre.
(Note that the initial upsurge in numbers is less marked with spaced point quadrats than contiguous area quadrats.)

In Britain we have adopted a simple system of classification in which vegetation types are identified by their dominant species. Thus, if alder continued to be dominant in the streamside sites we would accept them as part of the normal heterogeneity of an alder wood. However, if willows (*Salix* spp.) replaced alder as dominant, we would consider naming a second vegetation type: the decision to do so would require that the willow-dominated streamside areas be reasonably extensive. We have belonged to what may be called the dominance school, taking the view that the dominant largely determines conditions within the vegetation structure. To that extent the dominant determines which subordinate species may coexist with it. By analogy with taxonomy, the area dominated by a species is equivalent to the individual organism, while the group of areas dominated by the same species, the **association**, is the equivalent of the species and is the basic unit for classification. The association is named simply after its dominant: occasionally more than one species share dominance and feature in the association name.

In the rest of Europe little emphasis is placed on dominance, more on the social groupings of species. Indeed most *phytosociologists* make the point that common species (of which the dominant must be one) have broad ecological amplitudes—they are common because they can tolerate a wide range of environments. The rarer species, with narrow ecological amplitude, must be the best indicators of specific environments. However, their value for this purpose is largely negated by their rarity: they are unlikely to turn up in limited sample areas. Attention is therefore focussed on what they call the *character species* or *kenarten* of the vegetation type. These are the species which are at the same time reasonably common and yet reasonably exclusive to the vegetation type under consideration. Their basic unit for classification is also called the *association* which, as you may imagine, has been the source of much confusion. It is identified by the presence of a group of species of high *constancy* within the association which is much less consistently represented in other associations: these are the kenarten of the association and they may include inconspicuous herbs, mosses and liverworts. An association is only named when the vegetation type occupies an appreciable area on both a local and a regional scale. Thus a phytosociologist would be inclined to put the streamside areas of our alder wood into a separate association whether or not willow was dominant; but he would be influenced by their extent. Phytosociologists have built up a complicated hierarchical classification of associations, the groupings at each level being identified in the same way by species common within the whole group but relatively uncommon in other equivalent groups at that level. This is consistent but it does not tell us how we might set about defining the association level. Asking a phytosociologist what precisely constitutes an association elicits the same responses as when a taxonomist is asked to define the species. Apparently you must learn by experience rather than precept. We may note that the phytosociologist's association will frequently correspond to a sub-division of an association named by the dominance school.

Both of the above methods of classification have found wide application. The dominance method is particularly suitable for the initial work in areas where dominance by one or only a few species is the rule and the flora is not well known (limited taxonomic work required). The phytosociological method provides a more detailed classification of vegetation and has proved excellent as a basis for land capability mapping. However it demands a very complete knowledge of the flora. Outside Europe it has been applied most successfully in countries such as Canada where an adequate knowledge of the distribution of species in a not very rich flora could be rapidly built up. Neither method has proved very effective in the non-seasonal moist tropics where *poly-dominance* is the rule and we are still at the stage of accumulating data on the distribution of species in an exceedingly rich flora. Both methods achieve the semblance of order by ignoring problem areas when building up the classification: subsequently these are treated as intermediate or transitional between described associations.

Other ecologists have been more impressed by the fact that no two assemblages of species are identical. Thus the Wisconsin school points out (Bray and Curtis 1957) that no two species have exactly the same range of tolerance of environment—their global, regional and local dispersions are all different in some degree. The maximum development of a species will occur where the environment is optimal and its representation will decrease towards its limits of tolerance. In this context we must remember that experimental data on optima are often misleading: maximum development in the field is often associated with a physical environment that is distinctly sub-optimal —the species has been excluded from its preferendum by competition. The bluebell (*Endymion non-scriptus*) is a plant of woods although it achieves much greater productivity in open conditions in gardens. Virtually all green plants are 'light demanders' in this sense and it is more logical to categorise species according to their ability to tolerate increasing shade.

The local dispersion patterns of species thus overlap in varying degrees and within the overlap area there will be gradations in their performance. The balance of representation will be continuously changing from place to place as a direct result of the great variety of environmental gradients operating: examples are soil depth, moisture regime, grazing pressure, light intensity, nutrient availability. Now the rate of change in vegetation will vary with the steepness of the critical gradients in the area. You will have come across extensive areas with miniscule gradients and little or no vegetational change just as you will have noted areas where change seems to be continuous. How can two such divergent situations be encompassed by one system of analysis? The Wisconsin school suggest that we describe vegetational gradients rather than attempt classification. Their main analytical technique is **ordination** by which they look for evidence of environmental gradients in the ranking of sample stands according to floristic similarity (see p. 114 for details). There is a tendency for ordination to suggest greater continuity in variation than actually exists on the ground. Bearing that in mind it is a most useful analytical tool, not only suggesting which are the more important environmental gradients but also indicating quite clearly situations that are amenable to classification.

Ordination approaches environmental gradients indirectly through vegetation gradients. **Gradient analysis** (Whittaker 1967) is the reverse, the study of vegetation in relation to recognisable environmental gradients on the ground. It is important not to confuse topographical gradients with environmental gradients: the two will often coincide, as, for example, the increase in soil moisture towards the bottom of a concave slope or the decrease in level of various temperature parameters with altitude. But a long even slope may only have marked environmental gradients near the top and bottom and some of the steepest environmental gradients occur on nearly flat land where the water table is near the surface. Generally gradients are steep between wet-land and dry-land and the change in character of the associated vegetation is correspondingly abrupt. When geological faulting or igneous intrusions bring into juxtaposition at the surface two very different rock types, factor gradients will also be very steep; the carboniferous-limestone and millstone-grit of the Pennines provide many examples—so, too, do basalt and limestone intrusions in the granites of the Scottish Highlands.

After a world-wide study of vegetation and its literature Webb (1954) concluded that the organisation of vegetation hovered tantalisingly between the continuous and the discontinuous. In particular countries the balance may swing one way or the other. A developed country like Britain has little of its original woodland:

what remains is fragmented and much modified. Often it has been managed in ways that have led to artificial uniformity in the tree layer. Planted woods have usually had to be protected in their early stages and are clearly delimited. Most have been monocultures. Thus there is no difficulty in identifying stands and their dominant species —a classificatory unit exists. The trouble is that the pattern of planting of indigenous species has not necessarily followed the pattern of the pre-existing woodland. This applies on both a local and a regional scale. Beech is virtually an exotic in Scotland. Pedunculate oak has undoubtedly been planted on many sessile oak sites in the north and west. And, of course, woods of exotic species are now far more extensive than those of indigenous species. The groundflora of a wood is likely to be a more reliable guide to site characteristics than the tree layer. Equally the groundflora is more sensitive to biotic factors such as grazing or recreational use. These are some of the reasons why a conventional woodland type approach will not work when we consider the whole ecosystem. The nearest to it here is the condensed account of woodlands by major dominant species, indigenous and exotic, which you will find in Appendix II. If a classification of woodland types is to convey total ecosystem information, it should not give undue weight to the dominant tree species. This suggests a sociological approach. Alternatively ordination methods may be preferred. Both require the compilation of species lists.

Practical work—woodland types

Students tend to arrive in sixth forms or at university with less and less experience in plant identification. They, and sometimes their teachers as well, are ill equipped to carry out floristic work of scientific value. The method below, based on phytosociological practice, has been devised with these limitations in mind. It requires accurate identification of only the common species; the others must be recognised as species but need not be named. A specimen may need to be taken of some common species for an identification check later. This emphasis on common species makes the method compatible with our current attempts to conserve our flora. In the past when field studies focussed special attention on rare species it was not unknown for a local population of such a species to be virtually eliminated as each unthinking student took a specimen.

1. *Field data collection and compilation of summary*

Your first concern in this work, if you wish to generalise from its results, is to ensure that any samples you assess are as representative as possible. The choice of a general location in the wood must be a subjective decision taken after a preliminary reconnaissance. Additional samples can be taken elsewhere in the wood as a check. Then there is the question of sample size and structure. A quadrat size of two metres square has been found suitable for use in many woodland types but it is instructive to investigate the species/area relationship yourselves and from it determine an appropriate quadrat size. Figure 10.2 shows a possible layout.

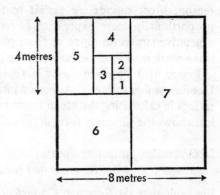

Fig. 10.2. A layout for investigation of the species/area relationship.

Groups of students in numbers proportionate to the area they have to search record all the species they can find in each subdivision. A further two areas would take the total to 256 m².

The central feature of the method is an estimate of constancy. This requires that there be several quadrats at least, the constancy of a species being estimated from the proportion of the quadrats in which it appears. A minimum of ten and a maximum of twenty quadrats is recommended. Supervision is easier if the quadrats are not too far apart. Throw a stick over your shoulder to decide the starting point: lay out a metre tape from that point into the area you have chosen: place your quadrats at five metre intervals on alternate sides of the tape. Each student group needs to be equipped with four meat skewers joined by two metre lengths of string. Now record all species present and their importance value on the Domin Scale (Chapter 2, p. 24). Remember to note the tree and shrub canopy above the quadrat in the same way—species and completeness of cover. The lists may be limited to flowering plants and gymnosperms or they may include, progressively, ferns, mosses, liverworts and lichens.

Ability to recognise the common species of the group is the criterion for their inclusion. Only those species present in 60 % of the quadrats will be recorded in the summary and these must be accurately identified. A similar criterion applies to groups below the level at which you have decided to record. Take a specimen of any such species which achieves 60 % constancy, have it identified subsequently and record it in the notes at the foot of the summary. The supervisor must co-ordinate the codes used for species which are not known e.g. *Carex* sp. B or moss C etc., and he must take the responsibility for collecting a specimen when this is necessary.

A problem which crops up in all ecological work of this kind is the change in appearance of individuals of a species through the season. Most species are easiest to identify in mid summer and proportionally more experience is required in identification from vegetative structures alone or from plants with dehisced fruits etc., if the work is to be done earlier or later.

Figure 10.3 opposite shows a typical sample compilation with Domin number ratings transformed into Bannister cover/abundance ratings in calculating the Mean Importance Value. Figure 10.4 overleaf shows the summary form derived from it.

2. *Ordination—assessing the similarity of samples and displaying the affinities of related samples.*

First calculate the Sørensen *Coefficient of Similarity* (*S*) between all possible pairs of species lists

$$S = \frac{2c \times 100}{a+b}$$

where a and b are the numbers of species in the two lists being compared and c is the number of species shared. The lists can be those for individual quadrats or those for a series of quadrats combined as in Fig. 10.2 above. Obviously all lists compared must have been collected in the same way, which argues for the standardisation of quadrat size and layout. The degree of similarity to be expected between different samples from what appears to be a homogeneous area can be determined by calculating similarity coefficients between different quadrats in the same sample. The average value for *S* between the quadrats detailed in Fig. 10.3 is

Fig. 10.3. Sample data from a sessile oakwood in Argyll 5 July, 1971.

Quadrat No.	1	2	3	4	5	6	7	8	9	10	Constancy	Mean Importance Value
		2		4		6		8		10		
	1		3		5		7		9			
DOMIN SCALE NUMBERS												
Trees												
Abies grandis(s)				+		+		+		+	4	<1
Fagus sylvatica(s)		1			1	+			+	+	5	<1
Picea sitchensis(s)								1			1	<1
Quercus petraea	8	9	9	9	9	9	9	9	9	5	10	68
Sorbus aucuparia(s)				+			+	1			3	<1
Thuja plicata(s)									1		1	<1
Shrubs												
Calluna vulgaris						2					1	<1
Lonicera periclymenum	3			+	+	+	2			3	6	2
Salix sp. A	3										1	<1
Vaccinium myrtillus		6	+	6	5	5	3	4	4		8	20
Dicot Herbs												
Galium saxatile		2		2		2	3		2	2	6	3
Lysimachia nemorum	3		1						+		3	1
Oxalis acetosella	4	4	4	3	4	4	5	3	3	3	10	20
Potentilla erecta		+			2	+	+				5	1
Viola riviniana									3		1	<1
Primula vulgaris	3		3						+		3	2
Rumex acetosa	3										1	<1
Grasses												
Agrostis canina	3		3				3	2			4	3
A. tenuis		4			4	4					3	8
Anthoxanthum odoratum	2		4		1		3			2	5	5
Deschampsia flexuosa		7	3	7	3	3	6	5	4	4	9	24
Festuca ovina	+										1	<1
Holcus mollis						+	+				2	<1
Molinia caerulea								+			1	<1
Other Monocots												
Carex pallescens	2										1	<1
Endymion non-scriptus						2	3			2	3	2
Luzula pilosa			+								1	<1
L. sylvatica			2			2	2	2		3	5	3
Ferns												
Blechnum spicant	2		+	2	2	3	3	3	1	2	9	5
Gymnocarpium dryopteris	3		4		+		2			3	5	5
Thelypteris limbosperma			1				+	1			4	1
Pteridium aquilinum							+		3	1	3	1
Mosses												
Dicranum majus	3	5	3	5	5	3		4	4	3	9	18
Hylocomium splendens				5	3	4		3	3	2	6	9
Hypnum cupressiforme	5	2	4	6	4	2					6	13
Mnium hornum		2									1	<1
Plagiothecium undulatum	3	2	2			3		2			5	3
Pleurozium schreberi		3				4	2	2	2		5	5
Polytrichum formosum	3	3	2	3		2	4	4	4	3	9	12
Pseudoscleropodium purum				2							1	<1
Rhytidiadelphus loreus	3	3	2	7	3	3	3	2	2	2	10	11
Thuidium tamariscinum	4	2	5		2	4	4	2	3	4	9	16
Thamnium alepecurum		5		4				2			3	6
Moss sp. A.		2									1	<1

Fig. 10.4. The sample summary derived from data in Fig. 10.3.

VEGETATION TYPE Western Oakwood

Location	Ballantyre Wood, Argyll Estates, Inverary M.R. 033116	**44**
Site	Steep east-facing slope with some rock and block scree.	**Total Species Recorded**
Sampling Method	$10 \times 4m^2$ quadrats in traverse at 5m interval.	

CONSTANT SPECIES (60%+ underlined 80%+)

		Mean I.V.
TREES (6)	*Quercus petraea* (sessile oak)	68
	(all other tree species as seedlings)	
SHRUBS (4)	*Lonicera periclymenum* (honeysuckle)	2
	Vaccinium myrtillus (bilberry)	20
DICOT HERBS (7)	*Oxalis acetosella* (wood sorrel)	20
	Galium saxatile (heath bedstraw)	3
GRASSES (7)	*Deschampsia flexuosa* (wavy hair grass)	24
Other MONOCOTS (4)		
FERNS (4)	*Blechnum spicant* (hard fern)	5
Other FORBS (0)		
MOSSES (12)	*Dicranum majus*	18
	Hylocomium splendens	9
	Hypnum cupressiforme	13
	Polytrichum formosum	12
	Rhytidiadelphus loreus	11
	Thuidium tamariscinum	16

NOTES Oakwood grown up from coppice, perhaps 140 years old. Gaps opened and NW American conifers planted in groups in the 1880s (hence the conifer seedlings). A fairly characteristic wet western oakwood with filmy fern present (*Hymenophyllum*), oak fern (*Gymnocarpium*) and mountain fern (*Thelypteris*). There were many leafy liverworts not recorded here—*Bazzania trilobata* was abundant with *Thamnium* on the boulders and *Frullania tamariscinum* on the stem bark of the oaks. Bluebells and primroses were present. The *Polytrichum* recorded is almost certainly *P. formosum* but no microscopic check was made.

59·5 (range 42–77). Enter the S values in a matrix as shown in one half of Fig. 10.5 below. Now subtract each of these values from 100 to obtain dissimilarity or *distance* indices and enter them in a second matrix. The next stage is easier if the distance values are duplicated to fill the whole of the square. Search for the highest distance values in the matrix. If there are one or more values of 100 (no species in common) then the range of community types is too great for this kind of presentation. The two plots with the highest distance value are to be used as the first axis in a double ordination of the samples. All the other samples are now placed in order along this first axis according to their distance from the two axis samples.

Coefficients of similarity

quadrat	1	2	3	4	5	6	7	8	9	10
1	+	43	70	53	49	44	55	42	44	57
2	57	+	56	61	56	59	56	54	69	54
3	30	44	+	55	61	60	67	54	57	68
4	47	39	45	+	61	70	56	71	63	53
5	51	44	39	39	+	75	56	54	63	63
6	56	41	40	30	25	+	70	63	67	71
7	45	44	33	44	44	30	+	55	63	77
8	58	46	46	29	46	37	45	+	67	52
9	56	31	43	37	37	33	37	33	+	65
10	43	46	32	47	37	29	23	48	35	+

Distance values

Fig. 10.5. A matrix of coefficients of similarity and the distance values derived from them. Data from Fig. 10.3 is used.

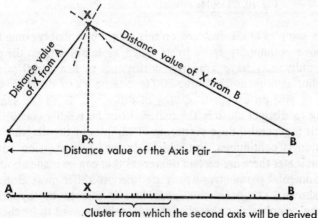

Fig. 10.6. Geometrical ordination.

This is done as in Fig. 10.6, using a geometrical method, though P_xB can be calculated also from the formula:

$$P_xB = AB/2 - \frac{(AX)^2 - (BX)^2}{2\,AB}$$

The ordination that results (lower part of Fig. 10.6) usually shows some clustering of samples at, or not far from, the centre of the axis. Mark these cluster samples on your matrix and find out which pair of samples among them has the greatest distance value. Repeat the ordination procedure for all other samples on this second axis.

You now have, for each sample, a position on two axes and you can show their positions relative to each other on a *scatter diagram*, Fig. 10.7.

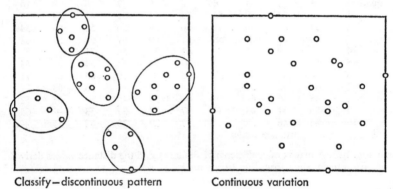

Classify—discontinuous pattern   Continuous variation

Fig. 10.7. Double ordinations (scatter diagrams).

Any samples at all clustered on this diagram probably come from the same community type. In as much as samples from the same community may have similarity coefficients as low as 60 (species rich) but seldom much above 80 (species poor) so distance value tends to give an exaggerated idea of difference. If all the samples belong to distinct clusters the material can be readily classified. If there is a general diffuse scattering of samples in the diagram then variation is continuous. But whichever type of dispersion pattern predominates there are certain inferences that can be made about the environmental parameters which are likely to differ most along the axes shown. The ordination procedure is said to be 'hypothesis forming', emphasizing that the inferences drawn need to be checked by measurement on the ground. If wetland samples appear in the middle and drier-land samples on either side, the axis gradient is not related to soil moisture regime. If the more diverse samples all appear at one side then there may be a nutrient gradient of some kind. Equally you can argue from your knowledge of the autecology of some species. For example, when rushes grow in what appears to be dry land vegetation you would suspect that the samples in which they appeared were wetter than the others. Make copies of your scatter-diagram by pricking through each sample point to pieces of paper below. Use one copy for each important species, marking in with circles those samples in which the species occurs: make the circles large when the species is abundant and smaller when the importance values are lower. These diagrams can be used both ways —to infer attributes of the axis gradient from a knowledge of individual species or to infer species attributes from a knowledge of the likely axis gradient. Initially, look up the habitat notes for each important species in a flora and deduce the likely environmental parameter gradients in that way. Remember that there is no law which says that gradients determined in this way are caused by only one environmental factor. Sometimes it is evident that two major factors are interacting when the scatter diagram trend is diagonally across. Aeration and dryness are attributes that often maximise together.

Some environmental attributes can be estimated without sophisticated apparatus or techniques. Altitude and topographical situation may be used as a guide to the temperature regime, frostiness and exposure. Rainfall isohyets, however, are a rather poor guide to local precipitation which can vary considerably over distances of only a few miles. Soil moisture will vary with the recent incidence of wet weather, but it may be possible to categorise relative wetness in a series of samples taken on the same day. Soils can be ranked for depth, stoniness, parent material, organic matter content or acidity (pH). If you have the equipment to make more precise environmental parameter measurements, so much the better.

Plot the values (categories) of each parameter you have estimated on a copy of the scatter-diagram (pricked through onto tracing paper in this instance). Try to draw in iso-lines or boundaries between categories. Superimpose the tracing paper with its iso-lines on each of the species diagrams in turn. How good is the correlation? With quantitative data you could calculate the *correlation coefficient* **r**:

$$r = \frac{\Sigma[(x-\bar{x})(y-\bar{y})]}{\sqrt{\Sigma(x-\bar{x})^2 \Sigma(y-\bar{y})^2}}$$

where x, y are the values of the environmental parameter and species importance and $\bar{x}$, $\bar{y}$ are the means of these values. You are not likely to get values as high as with the regression of tree dry weight on girth, but values above 0·4 (+ or −) may be significant if there are many samples. The suggestion is then that the factor is either responsible in part for the species dispersion pattern or that the factor itself is strongly correlated with a causal factor.

3. Classification

Throughout this book the emphasis has been on the important role which non-professional ecologists can play in increasing our understanding of woodlands. As you accumulate sample data make sure that you preserve at least the summary sheets as a permanent record. From them, over the years, you will build up enough material to make practical classifications of local woodlands. In woodlands where you have repeated work annually you will have a record of change with time. If you can get other ecological classes to standardise on method, then you can exchange data and build up a local reference collection of sample summaries that much more quickly. When teachers change the new incumbent will be able to familiarise himself or herself with the local woodland that much more readily.

Appendix I What are ecosystems?

In 1877 Mobius published a classic paper about an oyster bed describing it as a *biocoenose* to stress the way plants and animals interact with each other in the community. Tansley, a botanist, introduced the concept to Britain translating biocenose as **ecosystem**. In so doing he subtly transformed the concept and initiated the approach to ecology that has been developed so extensively in the last few decades. The term ecosystem automatically suggests that there is *system* in the way plants and animals react to one another and to the physical environment.

Tansley was probably thinking of vegetation units and mentally adding in the animal component. Indeed, among terrestrial ecosystems, this is often justified; for communities of animals and microorganisms are mainly co-extensive with the plants upon which they depend for food or shelter. Today, the term ecosystem is used much more loosely for any community whenever emphasis is to be placed on the functional relationships between its species populations. The plant-animal communities developing in puddles have been said to constitute an ecosystem. The term is even used for the largely invisible community of bacteria, yeasts and filamentous fungi that develops on most leaf surfaces.

It seems more logical, however, to treat these 'ecosystems-in-miniature' as sub-systems within ecosystems on the scale that Tansley envisaged. There will often be difficulty in deciding precisely where the boundaries of an ecosystem are located. It is the same problem as determining where vegetation boundaries should fall (see Chapter 10). The main structure of a terrestrial ecosystem is created by the larger plants present. Their above-ground parts and roots will often define the vertical limits of the ecosystem. We shall be interested in that structure according to the extent to which it determines how an ecosystem functions.

The complexity of massive present-day ecosystems is often

bewildering. The essence of the system is perhaps most clearly seen when considering the stage in the evolution of life on this planet when an ecosystem first came into being.

We know now that many organic compounds can be synthesised in nature without the intervention of living organisms. The first life forms are postulated as arising in shallow pools of organic-matter 'soup'. Evaporation would lead to a necessary concentration and sometimes to a complete drying out, when material may have been blown about by wind. Individual life forms may have gained an identity through desiccation: certainly the early life forms were capable of withstanding desiccation and wind dispersal to other pools. If the pool colonised contained the right kinds of organic matter, the life-form became active again. Imagine that the main food (energy source) of this life-form is organic compound A which is degraded only as far as organic compound B (which must therefore have a lower energy content). Now imagine a second life-form using organic compound B as its food, degrading it to inorganic substances, and we have the essentials of an ecosystem.

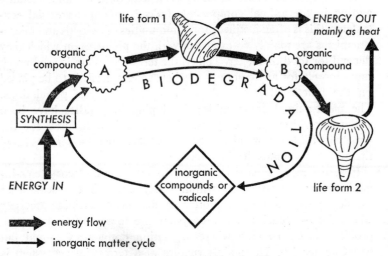

Fig. AI.1. The first "Ecosystem"

Note that the system involves

1. the complete degradation of organic compound A to inorganic matter with the release of energy
2. one functional relationship between life forms, the second being dependent on the first for the production of compound B.

The system is dependent on the accumulation of substance A in the first place and dispersal mechanisms efficient enough to ensure that both species arrive at sites of such accumulation at regular intervals. With the evolution of photosysthetic organisms, perhaps many millions of years later, the potential for biosynthesis of organic matter increased by many orders of magnitude and abiotic synthesis of organic compounds became relatively unimportant as an energy source. In ecological jargon, the first ecosystems were **hetero-trophic** (dependent on organic matter supplied) while most ecosystems today depend on the ability of **autotrophic** (green) plants to trap radiant energy.

There seems to have been a principle operating: that whenever an organic matter store built up in an environment not too severe to support life, a life-form would eventually evolve to exploit it. Organisms are no doubt evolving at this very moment in response to the stores of polythene and other new polymers that we scatter about the world. This principle applies not only to specific organic substances which we can lump as **dead organic matter** (D.O.M.) but also to each new species as it becomes abundant. Whenever the numbers of a species built up in sufficient quantity over a long enough period life forms evolved that could exploit them as energy sources. *Herbivores* and *parasites* evolved capable of exploiting plants. *Carnivores* and *hyperparasites* followed them.

The evolution of plant and animal species led to increasing size and structural complexity in both the individual and the ecosystem but the fundamental processes remained the same:

1. the **synthesis** of organic matter, now predominantly the bio-synthesis of new cell material (*biomass*), the energy deriving in the first instance from the light rays of the sun.

2. the **biodegradation** of biomass and D.O.M. derived from it, with energy loss as heat and the release of inorganic chemicals.

The whole system can be quantified in the same energy units, though radiation income, the calorific equivalents of biomass and

E

D.O.M., and heat loss during respiration are normally recorded in different energy units. Because energy cannot be lost, only transformed, energy budgets must balance. Use may sometimes be made of this fact to impute the activity of certain groups of organisms for which direct methods of investigation are not yet available. Using the *energetics* approach, energy is seen as flowing into the ecosystem, flowing between various organic matter stores and flowing out as heat. Energy flow through the system is, however, only a by-product of the two main processes described above.

Biodegradation, when complete, sees the release of the inorganic chemicals required for further biosynthesis by autotrophs—a literal *feed-back* process, which results in **nutrient cycling**, a circulation of elements within the system. If the losses and gains of nutrient elements in an ecosystem balance, then the system is potentially capable of functioning as long as energy inflow continues.

Ecosystem ecology is thus concerned with the details of synthesis, or production, and biodegradation, or decomposition. As these processes result from the activities of the component populations, ecosystem ecology is concerned with the linking up of population activities, the study of their interrelationships. The role of populations in what we may call ecosystem metabolism is taken up in Chapter 5.

Appendix II What to do and where to do it

This little book has been largely concerned with the recognition and measurement of change in woodlands—with the distinction between mere fluctuations in the amplitude and structure of component populations on the one hand, and cyclic or directional changes on the other—with the identification of directional change due to extrinsic factors. Over short periods the changes are small. Careful, painstaking, practical work is essential if they are to be identified. All the practical work suggested here will become more meaningful if set in the context of continuing (open-ended) investigations of one area. Successive years' work will reinforce early results: aberrant data will come to be recognised as such. Obviously the work will be seen to be most useful when you are able to confirm or refute suspicions that certain extrinsic factors are leading to system degradation, the lowering of production, the reduction of diversity. There must be strong incentives to study a woodland area which you believe to be under pressure from human activity, either directly or indirectly. Common examples will be woods of indigenous species under grazing or recreational use pressures. Equally important questions arise in woodland dominated by exotic species. How is the new dominant affecting the physical environment, especially the soil, through litter breakdown characteristics? Are the local fauna and flora able to adapt to this regime and so produce an ecosystem that functions as well as its indigenous predecessor? These two categories of woodland are commonly encountered, another cogent argument for working in them.

Woodland owners, be they private individuals, Corporations or the Forestry Commission, will generally give sympathetic consideration to requests for access to do ecological work. Ask whether there is a management plan covering the area in which you want to work, as then there are likely to be records of past history, of work done and production obtained.

125

In the green-belts of towns and cities there may be planting plans but it will be unusual to find a management plan covering the older woodland areas. This is hardly surprising when you consider that the management objective would be amenity rather than wood production. Amenity is a rather intangible asset that people tend to take for granted until threatened with its loss. Only by the slow process of civic education can the public be brought round to the conservationist point of view. Many of the woods in city confines were planted 150–200 years ago and have existed virtually untended for the greater part of that time. They cannot be preserved indefinitely but their amenity can be conserved by the very gradual replacement of over-mature individuals and parts of over-mature stands. Ecological work in green-belt woodlands will begin to provide data for management as well as opportunities to bring the problems of town woodlands to the attention of the public.

Outside the towns and cities different kinds of woodland are also associated with distinctive periods of our forest history and therefore tend to show a limited age range. There are notable exceptions to this rule. The pine woodlands in the Spey valley around Aviemore (the old Rothiemurchus Forest) regenerate naturally and freely after felling: there are stands of all ages up to maturity, though the forester's concept of maturity is now such that few stands are left to grow beyond 120 years. The beech woodlands of the Chilterns also show the full age range, though regeneration is commonly by planting. More often management, even when subscribing to the principle of sustained yield, followed the fashion of the day, planting the species that seemed at the time likely to yield a good financial reward in due course.

Exotic species have always had their fascination. In the eighteenth century these were mainly species from Europe, larch, spruce, silver fir and various pines. In the early nineteenth century the great flood of introductions from all over the world began. The ones that flourished best came predominantly from the northwest coast of North America, a region with a climate not too unlike our own. Many of them came to be grown in small plantations or planted in groups in broad-leaved woodland during the latter part of the century.

The first world war showed that our woodlands could not sustain us, even for a few years, when timber imports were cut off and the Forestry Commission was formed to re-establish our forest resources on an adequate scale. The land they were able to acquire was mainly land marginally suitable for agriculture, of low productive potential and often in the hills or mountains. It took several decades to develop successful, cheap, large-scale techniques of re-afforestation and the accumulated experience of exotic conifers on old forest land was of limited value. Thus it happened that Forestry Commission plantings only burgeoned after the second world war. Two-thirds of the area planted by 1973 consists of stands less than 25 years old. Sitka spruce, which does so well on the drained blanket peat of the north and west of Britain, is the predominant species with lodgepole pine increasingly used at high elevations and in nutrient poor sites. The proportion of hardwoods planted in any year has usually been less than one per cent.

There follow brief notes on the major tree species of British woodlands, indigenous and exotic, and the kind of sites with which they have become associated. You have only to consult a standard reference work such as Tansley's *The British Isles and their Vegetation* to appreciate that the indigenous tree species occur over a range of soil types. Furthermore species ranges overlap in time (succession) and space. As a result there is a range of groundflora communities associated with any particular tree dominant. In some instances the groundflora communities in two stands dominated by different tree species may be more similar than communities found in different parts of the range of one species. That this is so is a reflection of the poverty of the British flora as compared with the rest of Europe. The phenomenon becomes more pronounced as you move northwards, even in Britain. In Finland, for example, Cajander (1909) found it possible to ignore tree cover and relate groundflora communities directly to soil and physiographic factors. Using the groundflora data alone he could predict which of the three major tree species (spruce, pine or birch) would be the dominant and identify plantations of one of these species which had replaced a prior stand of one of the other two species. This is the extreme situation, though such situations can sometimes be recognised in Britain through the same line of reasoning. The point being made is that there is no such thing as a typical groundflora community in oakwoods or pinewoods on a country-wide scale: though on a local scale with a restricted range of sites there may well be.

Another point that should perhaps be made is that the establishment of woodland dominants in the past required more than the ability to grow taller and far more rapidly than their competitors. The species had also to produce seed in sufficient quantity, well enough dispersed, to maintain that position. Thus the sweet chestnut (*Castanea sativa*) is obviously outside its natural range in Britain, just as beech is outside its range in Scotland—though it has produced good masts in the years following exceptionally sunny summers as far north as Morayshire and the northwest of Ross and Cromarty. Generally the area in which a species will succeed as a forest crop is much more extensive than that in which it is biologically capable of ensuring its survival.

1. Woods dominated by pioneer species

Alder (*Alnus glutinosa*)

On wet ground with a high water table for much of the year: soil a wet silty or clayey loam. Often pure and then almost always self sown. Common on alluvial flats, stream banks and slopes on which drainage is locally impeded as by the construction of a road along the contour. Dense, with very little ground flora in the building phase: a moderately rich groundflora in the mature phase (see Fig. 10.1).

A related species, the grey alder (*A. incana*), with much greater tolerance of dry conditions, has been planted extensively (sometimes with *A. glutinosa*) on mining spoil heaps in reclamation work. *A. glutinosa* is increasingly favoured for planting in mixtures on heavy clay soils because of the ability its roots have to penetrate clay and fix atmospheric nitrogen.

Birch (*Betula pendula* and *B. pubescens*)

Both are pioneer species with a wide range of tolerance. *B. pendula* (silver birch) extends more frequently on to dry, acid, sandy or gravelly soils. *B. pubescens* is commoner on wet sites in S.E. England and extends both higher in mountains and further north in Scotland. The latter species exhibits a wide range of leaf-form and growth-habit: several sub-species have been recognised at various times and hybrids with *B. pendula* occur (though they are not fertile and introgression does not develop). The canopy in a birch wood is never very dense and there is usually a moderately rich groundflora. However the range of soils on which birch can

Audic soils

occur is such that its groundflora communities can vary enormously—from communities with low diversity and heath or bog
affinities to communities with high diversity not very different
from those of the more fertile oakwoods. In the very north of
Scotland and at high altitudes *B. pubescens* woodland may
represent the climatic climax but at lower levels further south birch
woodland is usually seral to oak (or pine on low fertility sites in
the highlands). Such succession has often been prevented by
grazing pressure and over-mature stands of birch with a grass
dominated groundflora are a common sight in many parts of
Britain. Figure A II 1 overleaf summarizes the groundflora data
for a birch wood which has the potential for succession to dry oak
woodland.

Ash (*Fraxinus excelsior*)

Also a pioneer species, casting a light shade, but with rather high
nutrient requirements which it appears able to obtain in two
markedly different kinds of site. Ash woods are a feature of
shallow, often skeletal, soils over limestone and chalk, though
the moisture regime must not reach extremes of dryness in summer.
In the southeast and west to Monmouthshire such woods are
seral to beech. On the other hand ash is common beside streams
in the northwestern mountains on soils of only moderate base
status: ash appears to be able to obtain its nutrient requirements
when moving *eutrophic* water is available. Ash develops best
on the lower.colluvial soils of some depth found around the hill
massifs of base-rich rock. Here it is usually a co-dominant in
mixed oak woodland though it may locally form pure stands.
These are the only sites in which foresters will have planted ash
for timber production. Ash comes into leaf later than any other
British tree species and is among the earlier ones to lose their
leaves in autumn: its hardiness is thus due to avoidance of spring
and autumn frosts rather than tolerance of them.

Lodgepole pine (*Pinus contorta*)

A northwest American pine extensively planted over the last few
decades. There are many races and material from low elevations
should be called the beach or shore pine. It has the invaluable
combination of low nutrient requirements and the ability to with-

Fig. A II. 1

VEGETATION TYPE Birchwood

		41
Location	Speyside, Lagganlia MR 855037 c.270m a.s.l.	**Total**
Site	Morainic knoll, the east facing slopes	**Species**
Sampling Method	10 × 4m² quadrats in transect at 5m interval	**Recorded**

CONSTANT SPECIES (60%+ underlined 80%+)

		Mean I.V.
TREES (2)	*Betula pendula* (silver birch)	46
SHRUBS (3)	*Vaccinium myrtillus* (bilberry)	13
DICOT HERBS (18)	*Achillea millefolium* (yarrow, milfoil)	4
	Galium saxatile (heath bedstraw)	5
	Lathyrus montanus (bitter vetch)	3
	Potentilla erecta (tormentil)	7
	Succisa pratensis (devil's bit scabious)	5
	Viola canina (heath violet)	4
GRASSES (8)	*Anthoxanthum odoratum* (sweet vernal-grass)	15
	Festuca ovina (sheep's fescue)	36
	Deschampsia flexuosa (wavy hairgrass)	12
Other MONOCOTS (3)		
FERNS (0)		
Other FORBS (0)		
MOSSES (7)	*Hylocomium splendens*	22
	Pleurozium schreberi	15
	Rhytidiadelphus triquetrus	14

NOTES A small patch of naturally sown birch wood which may have appeared when morainic material was exposed during road making (perhaps 20 yrs old). A lot of lateral light and therefore very grassy. The diversity is high enough to suggest that succession to oakwood could occur if (a) grazing pressure is not too high and (b) there are parent oaks nearby (nearest at Kincraig across the valley).

June 1973

stand exposure. It is most likely to have been planted on si
have not carried woodland for a very long time—in the up
on poor sites and beside the sea in the west, again on the n
infertile sites.

2. Woods dominated by mid to late seral species

Beech (*Fagus sylvatica*)

Beech woods will be familiar to anyone living near the Cotswolds, the Chilterns or the North and South Downs, particularly as hangers, the woods on the steep scarp faces with their skeletal lime-rich soils. The shade cast by an individual large-crowned individual is so dense that nothing will grow directly beneath it. Groundflora is sparse and concentrated in or near gaps in the general canopy except for the bird's nest orchid (*Neottia nidus-avis*), the brown saprophyte (strictly a parasite on certain saprophytic fungi). Several beautiful Helleborine orchids (*Cephalanthera* spp. and *Epipactis sessilifolia*) occur with many other calcicolous species.

In the forester's sense beech matures in 150–180 years though the species has a much longer life-span and older individuals will sometimes be found in amenity areas. Except in the Chilterns there will probably be a big gap in the age classes available for study (corresponding to the period from the mid-nineteenth century to the second world war).

On the continent of Europe beech is largely a tree of mountains but such species often find the humid regime they need at lower levels in Britain's oceanic climate. There is every reason to believe that beech spread as far north as southern Yorkshire at one time and that, now, it could extend further into the north and west. There are planted beech woods throughout Britain but the sites they occupy are not necessarily those which beech would have come to dominate without Man's intervention. It is known that beech woods will grow quite well on even moderately acid sandy loams; however the result is usually litter accumulation, the development of mor humus and degradation of the soil. The species will not then regenerate naturally. You would have to judge for yourself the status of any northern or western beech wood you chose to study. The presence of herbs indicating reasonably high base status would indicate greater compatibility (e.g. dog's mercury (*Mercurialis perennis*) or wood sanicle (*Sanicula europaea*)).

Douglas fir (*Pseudotsuga menziesii*)

A northwest American species with an unusual taxonomic history having been placed in five different genera at various times. It was introduced by Douglas in 1827 (older names are *P. douglassii* and *P. taxifolia*) and for ninety years became increasingly popular as a plantation species. Then he Douglas fir chermes, a woolly aphid (*Adelges cooleyi*), decimated many stands especially of Colorado provenances (now separated as *P. glauca*). The many very fine stands that exist today are nearly all of coastal provenance. It thrives particularly on fertile well drained soils. However foresters have not continued to plant it on very fertile soils, partly because such land is seldom available for planting and partly because growth is then so fast that the timber becomes light and coarse grained. Plantations will usually be on ground of reasonable fertility, perhaps clayey but then on an appreciable slope. In the Lake District its leaves have been shown to carry nitrogen-fixing bacteria which continue to be active after the leaves have fallen to the ground. This may explain, in part, the beneficial effect on litter decomposition when Douglas fir is mixed with other species. It has often been planted on ground that used to carry oak coppice and it would be interesting to discover how local woodland plants are adapting to the micro-environment created by this exotic.

Larch (*Larix decidua, Larix leptolepis* and their hybrid)

These are mountain species of Europe and Japan introduced some 400 years ago and in 1861 respectively. They have been widely planted in hilly districts on soils derived from moderately to highly base rich formations, though they do not thrive on skeletal limestone soils. Very light shade is cast and a dense grassy ground vegetation may develop as the plantation is thinned. This eventually creates replanting problems and it has often been considered worthwhile to underplant with a shade tolerant species such as western hemlock (*Tsuga heterophylla*). It requires more soil moisture than Douglas fir but otherwise also tends to be found on what were oak coppice sites. Both species and the hybrid have been much used in lowland sites but then in mixtures and usually as a 'nurse' crop for another species (this means removal as poles after they have performed their function of controlling the ground vegetation during the first few years).

Oak (*Quercus* spp.)

Oaks provide a strong durable timber suitable for many kinds of construction, highgrade charcoal and tan bark. The few woods that have escaped intensive exploitation are found in rocky ravines and other places poorly accessible and of little agricultural potential. The predominant oak may be the sessile oak (*Quercus petraea*) or the pedunculate oak (*Q. robur*): from the Midlands northwards introgression in the sessile oak populations increases in frequency.

Oak woodland of the Atlantic period was called mixed oak forest because the pollen of elm (*Ulmus* spp.), lime (*Tilia* spp.) ash and cherry (*Prunus avium*) occurred in appreciable quantities with that of oak. We do not know to what extent composition varied from one site type to another. We do know that planting has taken place in virtually all the oakwoods that now remain; that management operations usually involved the systematic removal of less desirable species leading inevitably to more complete domination by oak.

It so happens that pedunculate oak has most of the attributes the tree planter desires in greater measure than sessile oak (larger acorns which store better and give a more robust seedling). Pedunculate oak has been consistently preferred for planting and huge quantities of its acorns have been imported from Europe. The late Professor Anderson went so far as to suggest that the pedunculate oak itself was an introduced species. Certainly it seems to have been introduced widely into the north and west of Britain on ground which is generally accepted to be the domain of the sessile oak.

Both species grow satisfactorily on a wide range of soil types of intermediate fertility. Both species can be the dominant in what have been distinguished as dry oakwood types on sandy rather acid loams and damp oakwood types on clayey near-neutral soils. It seems that only the pedunculate oak can make a fair showing on very heavy clay soils that are waterlogged for periods in winter while only the sessile oak is well adapted to the shallow, wet, but well drained soils of the palaeozoic massifs in the north and west. But, mainly because of widespread planting, the autecology of the two species in Britain is very imperfectly understood. The best starting point for a discussion of oak woodland types may thus be the past management system—whether it aimed to produce timber (the forester's high forest system), charcoal or tan bark (the coppice system) or a variety of products useful in a rural economy (coppice with standards system).

Woodland that has been managed for timber production

This will usually be dominated by pedunculate oak. Where both species occur together it may be, as Tansley suggests, that they are on ground equally suitable for both species—a deep well-drained but fertile sandy loam. The more likely explanation on other sites is that pedunculate oak has been planted on what was a sessile oak wood site and that natural seedlings of the latter have grown up with the planted seedlings. At least one example is known of a wood comprising what are commonly supposed to be first generation hybrids: the acorns are likely to have been collected from a few fine specimens of *Q. robur* standing close to a *Q. petraea* wood. Shrubs will usually be sparse and like the groundflora more closely allied to the type of soil and parent material than the dominant species of oak. The bluebell (*Endymion non-scriptus*), soft-grass (*Holcus mollis*), bracken (*Pteridium aquilinum*) community and its variants are found in dry oak woodland with either sessile or pedunculate oak as dominant. At the other end of the moisture gradient are damp oak woodlands on clay soils which may have a fair amount of hazel (*Corylus avellana*) undergrowth (a desirable species), primroses (*Primula vulgaris*) and wood sanicle (*Sanicula europaea*); on boulder clay with its lime admixture dog's mercury (*Mercurialis perennis*) forms dense societies and on soil rich in available phosphate the stinging nettle (*Urtica dioica*) may dominate the field layer; very wet clays may have the creeping buttercup (*Ranunculus repens*) as the dominant with rushes (*Juncus* spp.) and sedges (*Carex* spp.) in the permanently wet patches. There are many other familiar oakwood plants that occur over most of this range e.g. the wood anemone (*Anemone nemorosa*) and the bramble (*Rubus fruticosus* agg.) though the latter rarely becomes abundant in Scottish oakwoods.

Woodland derived from coppice

These are the woodlands that were managed for charcoal or tan bark production, mainly in the north and west of Britain, often remote areas but then accessible by waterway or from the sea. The dominant is usually sessile oak or introgressed sessile oak. The last regular coppice rotations were abandoned after the first world war and the woodlands that have grown up from coppice are now at least fifty years old (1974) and often well over a hundred years old. Multiple-stemmed trees are clear indication of coppice origin. However, the coppice stool stems may have been 'singled' and then there are only the rather crooked stems and poor general height growth as clues: if a stem has been felled recently you will see the characteristic coppice regrowth pattern—fast growth in the centre for some forty years and very close rings (slow growth) towards the outside. Occasionally there may be a maiden, a stem grown from a seedling, looking straighter and more vigorous and giving you some idea of the real site potential for oak. In Ireland there was a period of intensive exploitation of oakwoods by the iron-smelters but no subsequent management of the regrowth. With no planting of pedunculate oak to replace worn out stools, the Irish sessile oak woods are not introgressed at all.

Being mainly in the wetter, more oceanic parts of Britain these woods are rich in ferns and mosses (see Fig. 10.3). But they are on well-drained soils some of which can become very dry in periods without rain: here the bilberry or blaeberry (*Vaccinium myrtillus*) may be locally dominant with wood sorrel (*Oxalis acetosella*) in shadier patches and wood sage (*Teucrium scorodonia*) in open places. Here, too, is the wavy hairgrass (*Deschampsia flexuosa*) which may become dominant in open woods. On the most fertile soils at the foot of a slope or in hollows and ravines it is common to find that individuals of elm (*U. glabra*), ash, alder and sometimes sycamore have established themselves and the groundflora becomes very rich. Many species, such as hazel, primrose, wood sanicle and wood sedge (*Carex sylvatica*) are shared with the damp oakwood type in the Midlands.

Apart from holly (*Ilex aquifolium*) which appears in all the more oceanic oakwoods, large shrubs are usually sparse or absent. On the less fertile soils rowan (*Sorbus aucuparia*) and birch (*B.*

pubescens) may have become established as seedlings and grown through to the canopy where there were gaps.

Coppice with standards

Here the coppiced species is normally hazel while the standards are of pedunculate oak both maintained 'pure' when the woods were regularly cut. Standard trees and ageing coppice stools were replaced by planting and, indeed, some woods of this kind were actually established (or re-established?) on disforested land. The oak were widely scattered and for much of the short coppice rotation hazel did not cover the ground. Conditions were ideal for the luxuriant development of the herbaceous groundflora, especially in its vernal aspects. There may be 'sheets' of primroses, wood anemone and lesser celandine (*Ficaria verna*); species more characteristic of non-woodland areas (such as the daffodil (*Narcissus pseudonarcissus*) near the Forest of Dean) may also come in more freely and shrubs have many opportunities to become established. The soil is often a heavy clay and wet for much of the year.

This management system has largely lapsed. Sometimes the standards gradually disappeared as they became large enough to exploit, leaving hazel coppice with some self-sown trees and shrubs to grow up—oakwood no longer. Sometimes the area was retained as cover for game birds and occasionally oak was planted or conifers such as the Lawson cypress (*Chamaecyparis lawsoniana*). Many of the woods with mixed oak and conifers are areas of coppice oak or coppice with oak standards in which groups of exotic conifers were planted on an experimental basis during the latter half of the last century. The oak will be somewhat older. Some of these provide evidence for the effect of a variety of types of conifer litter on the soil forming processes and the ground vegetation.

Pine

Pine as we have remarked is a mid-seral species that can become the climax dominant on glacial sands and coarse gravels—also on north facing mountain slopes to over 650 m a.s.l. in Scotland. The remnants of such forests that have persisted in Scotland with relatively little change for most of the last 8 000 years have been described in detail in a book by Steven and Carlisle (1959). Pine

VEGETATION TYPE Mature Pinewood

Location	Speyside—NW of Lagganlia MR 853039
Site	Undulating (morainic) c. 270m a.s.l.
Sampling Method	Two transects A and B 50m apart: $11 \times 4m^2$ quadrats in each at 5m intervals

29

Total Species Recorded

CONSTANT SPECIES (60%+underlined 80%+)

	A	B		Mean I.V. A	B
TREES	(2)	(2)	*Pinus sylvestris* (Scots pine)	55	57
			Sorbus aucuparia (rowan)—seedlings only	4	7
SHRUBS	(4)	(3)	*Calluna vulgaris* (heather, ling)	4	9
			Vaccinium myrtillus (bilberry)	54	48
			V. vitis-idaea (cowberry)	24	18
DICOT HERBS	(5)	(5)			
GRASSES	(2)	(3)	*Deschampsia flexuosa* (wavy hairgrass)	16	28
Other MONOCOTS	(2)	(3)			
FERNS	(0)	(1)			
Other FORBS	(0)	(0)			
MOSSES	(8)	(8)	*Dicranum scoparium*	13	11
			Hylocomium splendens	44	45
			Pleurozium schreberi	15	24
			Rhytidiadelphus triquettus	30	27
			Plagiothecium undulatum (constant in B only)		10

NOTES Apart from the moss *Plagiothecium* the two transects produce the same list of constants with importance values (I.V.s) of the same order. *Goodyera repens* (creeping ladies tresses orchid) was present in both transects. Another orchid *Listera ovata* (lesser twayblade) was present in B and a large patch of the beautiful, rare *Linnaea borealis* occured in A. Note the presence as constants of low I.V. of heather and wavy hairgrass, one of which will dominate the site a few years after the wood is felled.   *June* 1971 colonises sandy areas in England freely but virtually all the extensive pine woodlands on sandy heathland in East Anglia and elsewhere have been planted. We have noted that even the unplanted pine in Scotland is more or less even-aged or two-aged. The groundflora is often dominated by heather (or ling, *Calluna vulgaris*) early in the regenerative phase—is virtually eliminated in the building phase—and is dominated by bilberry, cowberry (*Vaccinium vitis-idaea*) or robust hypnoid mosses (e.g. *Rhytidiadelphus* spp., *Hylocomium splendens* and *Pleurozium schreberi*) in the mature phase. The data in Fig. A II 2 were taken from a typical mature stand in the Rothiemurchus Forest area of the Spey valley. Note that heather and wavy hairgrass are constants though they have low importance values; either may become temporarily dominant after felling. Pine will colonise some wetter peaty areas with varying success. In the Black Wood of Rannoch, for example, the ground is very uneven: the pine has succeeded on knolls and mounds while bog plants such as purple moorgrass (*Molina caerulea*), cotton grass (*Eriophorum vaginatum*) and species of Sphagnum are found in the hollows between.

Spruce

Spruces probably dominate the largest area of woodland in Britain today. Both the common species are exotics. The Norway spruce, our Christmas tree (*Picea abies*), would probably have arrived via Scandanavia in due course without Man's aid. The Sitka spruce (*Picea sitchensis*) has been with us since the 1830's but the huge areas put down to this species mostly date from after the second world war. A pre-requisite for this development was the design of a plough big enough and robust enough to score deep drainage ditches through purple moorgrass bogs when the 'glacial till' beneath might include large boulders. Both species cast a dense shade in the pole stage (building phase). As these stands mature it will be interesting to see how their woodland flora and fauna evolves and what effect the combination of drainage and a conifer crop have had on the peat (or soil).

References

ANDERSON, J. M., 1973. Carbon dioxide evolution from two temperate deciduous woodland soils. *J. Appl. Ecol.*, **10**(2), 361–378.

ANDERSON, J. A. R., 1964. The structure and development of the peat swamps of Sarawak and Brunei. *J. Trop. Geog.*, **18**, 7–16.

ANDERSON, M. L., 1967. *A History of Scottish Forestry*, Oliver and Boyd.

ANDREWARTHA, H. G. and BIRCH, L. C., 1954. *The Distribution and Abundance of Animals*, University of Chicago Press.

BANNISTER, P., 1966. The use of subjective estimates of cover-abundance as the basis for ordination. *J. Ecol.*, **54**(3), 665–674.

BASKERVILLE, G. L., 1965. Dry matter production in immature balsam fir stands, *Forest Science Monographs* 9, Soc. of Am. Foresters.

BLACKMAN, G. E. and RUTTER, A. J., 1946. The interaction between light intensity and mineral nutrient supply in leaf development and in the Net Assimilation Rate of the bluebell (*Scilla non-scripta*). *Ann. Bot.* N.S. **12**, 1.

BRAY, J. R., 1964. Primary consumption in three forest canopies. *Ecol.* **45**, 165–167.

BRAY, J. R. and CURTIS, J. T., 1957. An ordination of the upland forest communities of southern Wisconsin. *Ecol. Monographs*, **27**, 325–349.

BRAY, J. R. and GORHAM, E., 1964. Litter production in forests of the world. *Adv. Ecol. Sci.*, **2**, 101–152.

BRIGHTMAN, F. H. and NICHOLSON, B. E., 1966. *Oxford Book of Flowerless Plants*, Oxford University Press.

BURGES, A., 1958. *Microorganisms in the soil*, Hutchinson.

CAJANDER, A. K., 1909. Über waldtypen. *Acta Forestalia Fenn.*, **1**, 1–175.

CARLISLE, A., BROWN, A. M. F. and WHITE, E. J., 1967. The nutrient content of tree stemflow and groundflora litter and leachates in a sessile oakwood. *J. Ecol.*, **55**(3), p. 621.

CLAPHAM, A. R., in Tansley, 1939. *The British Isles and their Vegetation*, Cambridge University Press.

CLAPHAM, A. R., TUTIN, T. G. and WARBURG, E. F., 1962. *Excursion Flora of the British Isles*, 2nd ed., Cambridge University Press.

CLOUDSLEY-THOMPSON, 1967. *Micro-ecology*, Studies in Biology 6, Arnold.

DE WIT, C. T., 1965. *Photosynthesis of Leaf Canopies*, Agricultural Research Report No. 663, Inst. for Biol. and Chem. Res. on Field Crops at Wageningen.

DIMBLEBY, G. W., 1952. The historical status of moorland in north-east Yorkshire. *New Phyt.*, **51**, 349–354.

ECOLOGICAL STUDIES, 1970. Ed. D. E. Reichle, Analysis of Temperate Forest Ecosystems.

ELTON, C., 1966. *Animal Ecology*, Science Paperbacks, Methuen.

FAIRBAIRN, W., 1972. Dalkeith Old Wood, *Scott. For.*, **26**(1).

FORESTRY COMMISSION PUBLICATIONS. F. C. Forest Record No. 58 is a check list of all publications between 1919 and 1965. F.C. Booklet No. 34; Forest Management Tables, G. J. Hamilton and J. M. Christie, 1971.

GODWIN, H., 1956. *The History of the British Flora*, Cambridge University Press.

HEWITT, C. G., 1921. The conservation of the wild life of Canada. New York. In: Elton, 1966.

HUDSON, H. J., 1972. *Fungal Saprophytism*, Studies in Biology 32, Arnold.

IBP HANDBOOKS—see individual authors, Blackwell Scientific Publications.

JACKSON, R. M. and RAW, F., 1966. *Life in the Soil*, Studies in Biology 2, Arnold.

JONES, E. W., 1945. The structure and reproduction of virgin forest of the North Temperate Zone, *New Phyt.*, **44**, 130–148.

KEBLE-MARTIN, W., 1965. *The Concise British Flora in Colour*, Sphere.

KERSHAW, 1964. *Quantitative and Dynamic Ecology*, Arnold.

KLOMP, H., 1966. The Dynamics of a field population of the Pine Looper, *Bupalus piniarius* L., *Adv. Ecol. Res.*, **3**, 207–305.

LAMB, H. H., 1972. *The Changing Climate*, University Paperback, Methuen.

LANGE, M. and HORA, F. B., 1963. *Mushrooms and Toadstools*, Collins.

LEBRUN, P., 1965. 'Contribution à l'étude ecologique des oribates de la litière dans une forêt de moyenne-Belgique', *Mem. Inst. Sc. Nat. Belg.*, **153**, 1–96.

MACFADYEN, A., 1963. *Animal Ecology* 2nd Ed. (Plates IIa and IIb), Pitman.

MACFADYEN, A., 1964. In: D. J. Crisp (ed.) British Ecological Society Symposium No. 4. *Grazing in a terrestrial and marine environment.*

MERTON, L. F. M., 1970. The history and status of the woodlands of the Derbyshire limestone, *J. of Ecol.* **58**(3), 723–744.

MOBIUS, K., 1877. In: E. J. Kormondy (ed.). *Readings in Ecology*, 1965, Prentice Hall.

NEWBOULD, P. J., 1967. *Methods for estimating the Primary Production of Forests*, IBP Handbook No. 2, Blackwell Scientific Publications.

ODUM, E. P., 1963. *Ecology*, Modern Biology Series, Holt, Rinehart and Winston.

ORLOV, A., 1955. The role of feeding roots of forest vegetation in enriching the soil with organic matter, *Pochvovedenie*, **6**, 14–20.

OVINGTON, J. D., 1957. Dry matter production by *Pinus Sylvestris* L., *Ann. Bot. Lond.* N.S. **21**, 287–314.

OVINGTON, J. D., 1962. Advances in Ecological Research I. In: J. B. Cragg, (ed.), *Quantitative Ecology and the Woodland Ecosystem Concept*, Academic Press.

PENNINGTON, W., 1969. *The History of British Vegetation*, English Universities Press.

PHILLIPSON, J., 1966. *Ecological Energetics*, Studies in Biology 1, Arnold.

POLLARD, E., 1973. Woodland relic hedges in Huntingdon and Peterborough. *J. of Ecol.*, **61**(2).

REICHLE, D. E., 1970. (ed.), Ecological Studies I. Analysis of Temperate Forest Ecosystems.

RICHARDS, O. W., 1926. Studies on the ecology of English heaths. *J. Ecol. XIV*, 244–281.

SHIMWELL, D. W., 1971. *The Description and Classification of Vegetation*, Sidgwick and Jackson.

SOUTHERN, H. N., 1959. Mortality and population control, *Ibis*, **101**, 429–436.

STEVEN, H. M. and CARLISLE, A., 1959. *The mature pinewoods of Scotland*, Oliver and Boyd.

TANSLEY, A G., 1939. *The British Isles and their Vegetation*, Cambridge University Press.

THOMAS, W. A., 1970. Weight and calcium losses from decomposing tree leaves on land and in water. *J. Appl. Ecol.*, 7(2),237–241.

TINBERGEN, L., 1960. The dynamics of insect and bird populations in pine woods, *Archs. néerl. zool.*, **13**, 259–472.

TITTENSOR, A., 1970. The red squirrel (*Sciurus vulgaris* L.) in relation to its food resource. Ph.D. Thesis, Edinburgh.

TITTENSOR, R. M., 1970. The History of Loch Lomond Oakwoods, *Scott. For.*, **24**, 100–118.

VEDEL, H. and LANGE, J., 1960. *Trees and Bushes in Wood and Hedgerow*, Methuen.

WALLWORK, J. A., 1970. *Ecology of Soil Animals*, McGraw-Hill.

WATT, A. S., 1947. Pattern and process in the plant community, *J. Ecol.*, **35**, 1–22.

WEBB, D. A., 1954. Is the classification of plant communities either possible or desirable? *Bot. Tidsskr.*, **51**, 362–370.

WHITTAKER, R. M., 1967. Gradient analysis of vegetation, *Biol. Reviews*, **42**, 207–264, Collier-Macmillan.

WHITTAKER, R. M., 1970. *Communities and Ecosystems*, Macmillan.

WYATT-SMITH, J., 1949. A note on tropical lowland evergreen rain forest in Malaya, *Malayan Forester*, XII, 58–64.

ZINKE, P. J., 1962. In: Kershaw, 1964, *Quantitative and Dynamic Ecology*, Arnold.

Glossary

Association—
    a) a vegetation type identified by its dominant species (Dominance School)
    b) a vegetation type of reasonable extent characterised by a particular combination of species of high constancy within it (Phytosociology)

Autecology—the ecology of a species (below the community level)

Autotrophic—a mode of nutrition with reliance on a physical energy source (not dependent on other organisms c.f. heterotrophic)

Basal area—the sum of the cross-sectional areas of tree stems measured at breast height (1.3 m)

Base line—a line laid out on the ground, used for locating sample units or transects

Biocoenose—the integrated plant/animal community

Biomass—weight of living matter

Bloomery—rural furnace, smelting iron ore with charcoal

Boreal—between Arctic and North Temperate zones

Calcicolous—associated with lime-rich soils

Character species—a species of high constancy in one vegetation type and one level of the phytosociologist's hierarchical classification

Climax—a relatively stable state reached by vegetational succession

Clitellum—a pronounced swelling covering several segments of a mature earthworm

C/N ratio—ratio of atomic carbon to atomic nitrogen in organic matter

Colluvial—receiving and losing sub-soil water in equal amounts c.f. Eluvial and Illuvial

Comminution—mechanical breakdown of organic matter into smaller pieces

Constancy—in phytosociology the proportion of vegetation samples in which a species appears

Coppice—woodland in which the woody species have been cut back to the ground and allowed to grow up again with multiple shoots

Constant—a species appearing in 80% of a set of vegetation samples

D.O.M.—dead organic matter

Deciduous—shedding all the leaves together

Density dependent—increasing *in intensity* with increasing density

Density related—rising or falling with density

Distance—an index of difference between vegetation samples derived from various indices of similarity

Diurnal—daily

Dominant—
    a) botanical—a species which by its size and/or abundance determines the environment enjoyed by associated species in vegetation: applicable to each stratum
    b) forestry—an individual tree with a pre-eminent position in the canopy
    c) ecosystem ecology—a species playing a major role in an ecosystem process

Domin Scale—a combined scale of plant cover and abundance (for species providing <5% cover)

Driving variable—an input to a system (systems analysis)

Ecological amplitude—the range of environments in which a species can survive

Ecological niche—see **niche**

Ecosystem—any reasonably well-defined area or space within which systematic relationships can be demonstrated between biota and between biota and the abiotic environment. Commonly used in a more restricted sense of units corresponding to individual terrestrial vegetation types

Ecotone—a transition zone between two vegetation types

Ectoparasite—see **parasite**

Edaphic—relating to the soil

Emergent—of a tree which extends above general canopy level—hence 'emergent stratum' as seen in coppice-with-standards

Endemic—of disease, at normal levels c.f. epidemic

Endoparasite—see **parasite**

Energetics—the study of energy flow in biological systems

Enumeration—the recording of numbers and usually other attributes of the individuals in a population (= census)

Environment—the totality of things, abiotic and biotic, around an individual, community or population

Epiphyte—a plant growing on an aerial part of another plant species

Epiphytotic/Epizootic—a disease or pest outbreak at epidemic level but affecting respectively plant and animal populations

Error (statistics)—the residuum of variability in a set of data after recorded sources of variation have been accounted for

Eutrophic—rich in nutrients, usually of soil water or a water body

Exotic—in forestry, a species not belonging to (indigenous in) a country

Extrinsic—originating outside

F-layer—see 'Litter layers'

Feed-back—a systems analysis term to describe an output which affects the operation of the system: positive feedback accelerates the system process; negative feedback slows it down

Food-chain—when the production of species A is the food of sp. B which provides the food for sp. C and so on

Food web—the normal situation with numerous cross links between identifiable food chains

Forb—a term devised to include ferns and herbaceous Angiosperms

Functional response—see **predation**

Glacial till—the ground-down rock material left when a glacier retreats

Gradient analysis—investigation of the way vegetation changes along recognisable environmental gradients

Gross production—see **production**

H-layer—see **litter layers**

Habitat—the place in which a species lives, the abiotic part of environment

Hangers—woods on steep escarpments, especially of beechwoods

Heterotrophic—a mode of nutrition with reliance on organic matter (and hence other organisms) as the energy source

Homeostasis—applied to systems in which negative feedback tends to damp down fluctuations away from the norm

Humus—partially decomposed dead organic matter

Hydrosere—primary succession beginning with open water

Hyperparasite—a parasite of a parasite

Infructescence—a plant structure bearing or including fruits

Interglacial—a major period of ice-retreat between successive glaciations

Interstadial—a minor period of greater warmth during a glaciation

Intraspecific—between individuals of the same species, c.f. interspecific

Intrinsic—originating within

Introgression—hybridisation followed by successive back-crosses with one of the parental species

Kenarten—character species in phytosociology

Lammas shoot—a second growth about the time of the Lammas Fair in August

Liane—a woody climber in tropical forests

Litter Layers—when the litter layer is thick and decomposition slow, three separate layers can be identified:
- a) the **L-layer** on top in which twigs, leaves, etc. are largely intact;
- b) the **F-layer**, the zone of active breakdown (fermentation) where macroscopic fragments are clearly recognisable; and
- c) the **H-layer** of fine organic matter particles just above the mineral horizon.

M.O.F.—mixed oak forest, the name for the woodland which gave rise to oak pollen deposits with variable amounts of pollen of elm, lime, ash and other species

Macrofauna—the larger animals (body size over 1 cm)

Maiden—a tree arising from a seedling in a wood of coppice origin

Mast—the fruit crop of beech, sometimes oak

Meiofauna—animals of intermediate size (= mesofauna)

Mesic—relating to the middle range of moisture regimes

Mesofauna—animals of intermediate size (200μ – 1 cm)

Microfauna—animals of body size less than 200μ (mainly Protozoa)

Microhabitat—a minute habitat

Mineralization—the process whereby mineral elements are released from dead organic matter into the ambient solution and become available again

Mor—aerobic brown-black humus lying above the mineral soil

Mull—paler or colourless aerobic humus incorporated *in* the mineral soil

Mutualism—the term preferred to 'symbiosis' for interspecific relationships in which both species derive advantage

Mycorrhiza—a rootlet modified in form by the presence in and around it of fungal hyphae, normally signifying mutualism (pseudomycorrhiza if the relationship is parasitic)

Net production—see **production**

Niche—a small place (as nesting niche), habitat with a specified characteristic (as brackish-water niche) and, the preferred use in this book, the functional place in an ecosystem (ecological niche, as indicated by such terms as herbivore or saprophyte)

Niche equivalence—when different species perform the same type of function

Numerical response—see **predation**

Nutrient capital—the sum total of nutrient elements retained in an ecosystem—normally applied only to elements that are sometimes in short supply

Ordination—the ranking of data sets, particularly of vegetation data and then an alternative type of analysis to classification

Palynology—the study of pollen and spores as fossils

Pannage—the practice of putting domestic pigs out to feed on fallen acorns and beech mast

Parameter—originally a constant in the formula defining a relationship—now widely used simply as a variable

Parasite—a species dependent on a live individual of another species (its host) for its food—ectoparasites live on the outside of their host, endoparasites within

Parasitoid—a species which lays its eggs in the tissue of its host, the host being killed eventually as the parasite develops within it

Periglacial—around the fringes of the ice-cap during a glaciation

Phenology—the study of climatic and especially seasonal effects on organisms

Phytophagous—herbivorous

Phytosociology—the study of the social organisation of plant communities within vegetation

Plagiosere—a succession diverted from its normal course by an extrinsic factor

Podsol—a soil type developing on siliceous parent material (acid and free draining) when litter forms mor humus above it: characteristically iron is leached downwards and may sometimes form an impervious layer or 'pan'

Point quadrat—a pin lowered into vegetation until it hits a plant or passes through gaps to the ground: the proportion of hits on a particular species estimates its relative cover.

Polydominance—dominance is shared by several species (at least) in space or time

Predation—the killing and subsequent eating of all or part of another animal species (the prey). Predation can be at *random encounter* level or it can be selective (*searching image* formed): when individual predators react to increased prey numbers by taking more per unit time, they show a *functional response*: if subsequently predator numbers increase, there is said to be a *numerical response*.

Preferendum—that part of the range of tolerance of a species in which it can be shown to function most successfully

Primary production—see **production**

Production—the biosynthesis of new organic matter (**Gross production**) some of which is required for maintenance (measured by respiration). **Net production** is Gross production less respiration losses. **Primary production** is the production of autotrophs while **Secondary production** is by heterotrophs

Productivity—a rate, the amount of production per unit time and, in ecosystem ecology, per unit area, e.g. tha^{-1} $year^{-1}$ or gm^{-2} $year^{-1}$

Profile diagram—a descriptive display technique for vegetation in which a representative vertical section is drawn

Provenance—the locality of origin of a seed-lot or other propagation material

Regeneration—young trees replacing or capable of replacing old ones (forestry)

Regression (statistics)—a trend relationship between two variables when one is a function of the other

Saprophyte—a plant using D.O.M. as its energy source

Saprovore—an animal consuming D.O.M.

Scatter diagram—a display technique possible whenever there are values for two parameters describing one set of entities, used for double ordinations of sets of vegetation data

Searching image—see **predation**

Sere—a succession of vegetation

Similarity coefficient—a mathematical formulation relating characteristics shared by two entities to the sum total of characteristics—in vegetation analysis relating species shared to the total number of species present

Simulation model—a mathematic model which attempts to reproduce the input/output relationships of natural systems

Species/Area curve—a plot of cumulative number of species encountered against the area enumerated

Stand—an area of more or less uniform vegetation selected for study

Standcycle—the cyclic changes in a stand consequent upon even-aged structure

Standard—individual trees grown for large timber among coppiced trees (Coppice-with-standards)

State variable—a variable store within the system (systems analysis)

Stratification (statistics)—when sampling, the division of a non-homogeneous entity into two or more relatively homogeneous parts each of which is sampled separately

Stool—the roots and basal part of a coppiced tree

Sub-sere—the succession resulting after partial destruction of vegetation

Symbiosis—two species living in close association with each other—originally confined to mutualism but now more appropriately used to include most parasitic relationships

Synusia—a category of plant defined according to the way it accomplished its functions as an autotroph or heterotroph, e.g. climbers

Systems ecology—the application of systems analysis to ecological systems

Thinning (forestry)—the selective removal of trees in a dense stand to improve the performance of the remaining individuals

Tree-line—the uppermost limits of tree growth on mountainsides

Trophic level—the number of times removed (along a foodchain) from autotrophs, which constitute trophic level 1

Turnover—that part of the accumulated production of a population or ecosystem which dies and is no longer reckoned part of its biomass

Zone-line—a line in rotting wood that marks the contact interface of two antagonistic fungal mycelia

Index

Age structure, population, 19, 21, 31, 33, 36, 44, 126
Amenity, 126
Annual rings, see Growth rings
Association (Dominance School), 109
— (Phytosociology), 109

Bannister cover/abundance scale, 23, 114
Basal area, **60**, 72, 106
Biomass, 5, **6**, 17, 20, 55, **Ch. 6**, 80, 85, 102
Black Wood of Rannoch, 137

Canopy strata, 7
Capture/recapture technique, 79, **89**, 91
Chiltern beechwoods, 126
Classification, of woodland, Ch. 10
Clearance, woodland, 34, 42, 45
Climate, 19
—, optimum, 42
—, post-glacial, 41–44
Climax, climatic, 31, 42, 129
—, diverted, 32
—, edaphic, 31
—, fire, 33
C/N ratio, 53–55
Coefficient of Similarity, 114
Comminution, 52, 53, 55, 79
Competition, 20, 26, 86, 110
Conservation Acts, woodland, 45
Constancy, 58, 109, 113, 115
Consumer, primary, 51, 54
Coppice, 34, 46, 48, 132, 135
Coppice with standards, 9, 138
Correlation coefficient, 73, 119
Cover/abundance scale, Domin, **23**, 133, 114
—, Bannister, **23**, 114

Decomposers, 51, Ch. 7

Decomposition, 7, 11, 19, 53, 55, 57, **Ch. 7**, 132
Defoliation, 87, 88, 104
Dilution plate technique, 77, 81
Diurnal change, 18
Diversity, 32, 40, 47, 57, 59, 119, 125
Domin scale, **23**, 113, 114
Dominance, 2, 11, 32, 109
Dominant, botanical, 11, 26, 109
—, ecological, 2, 11
—, forestry, 6, 11
Drainage, 42, 48, 137
Dry weight determination, tree, 68–72
Dynamic equilibrium, 18, 102, 105

Ecosystem, 2–4, 13–17, 19, 50, 51, 91, 93, **App. I**
— models, 99, 100
Ecotone, 28, 107
Enclosure Acts, 44
Energetics, 124
Energy flow, 58, 124
Enumeration, **14–16**, 22, 35, 47, 60–63
Epiphytotic, 87, 104, 106

Fire, 33
Food chain, 50, 93
— webs, 51, 58
Forest of Dean, 136
Forestry Commission, 1, 47, 61, 72, 81, 98, 101, 102, 125, 126
Fruit production, tree, see Seed production

Gap, 7, 17, 135
Gas analysis, micro-techniques, 79
Glaciations, 28
Gradient analysis, 111
Grazing, 30, 36, 42, 44, 45, 125, 129
Groundflora, 8, 10, 16, 22, 34, 58, 65, 68, 73, 89, 112, App. II
Growth rings, 35, 37, 62, 64, 135

147

Height, top, 72, 102
Herbivore, 19, 36, 55, **87–91**, 93, 94, 96, 99
Historical factors, Ch. 4
Humus, 7, 76, 131
Hydrosere, 27, 31

Ice Ages, 38
Ice Age, Little, 94
Interglacial periods, 38, 39
International Biological Programme, 3, 17, 64, 98, 102
Introgression, 128, 135

Reverse–J curves, 31, 32

Leaching, 30, 54, 57, 62
Life table, 104
Light intensity, 7, 8, 17, 18
Litter, 7, 11–13, 46, 78, 83, 131, 132, 136
— fall, 11, 13, 22, 62, 63, 65, 74, 75
— layers, 76, 77
Longevity, tree, 5

Main storey, 9
Management, woodland, 1, 3, 21, 34, 46, 126
Marginal land, 34, 44, 127
Micro-habitat, 11, 17, 49
Mineralization, **53**, 54
Models, ecosystem, Ch. 9
Mor humus, 76, 131
Mortality, 62, 96, 104, 106
Mull humus, 76
Mutualism, 95
Mycorrhizae, 53, 95

Nature Conservancy, 1
Niche, 41, Ch. 5, 59, 74
—, ecological, **49**, 50, 57
—, equivalence, 50
Nutrient capital, 26, 28, 55
—, cycle, 19, 55, 57, 95, 124
— status of soil, 28
— uptake, 19, 53–55, 96

Ordination, 111, 114–120

Palynology, 39
Parasitism, 55, **94–97**
Phenology, 58

Phytosociology, 109, 112
Pioneer species, trees, 26, 35, 128–131
Pit-fall traps, 79
Plagiosere, 30, 36
Planting, tree, 34, 46, 48, 89, App. II
Point quadrat, 24, 25
Polydominance, 110
Population activity, see Production
Population age structure, 19, 21, 31, 33, 36, 44, 126
— density, 106
— importance, Ch. 6–8
Predation, 51, 78, **91–94**, 96, 99
Pressler borer, 37, 62
Production, primary, 19, 55, Ch. 6, 87
—, secondary, 50, Ch. 7, Ch. 8.
— model, 99–102
Profile diagram, 9, 14

Quadrat, 22, 98, 113, 114
—, point, 24, 25

Regeneration, woodland, 2, 20, 22, 31, 33, 126
Regression, 61, 99
Respiration, 53, 61, 75, 79, 124
Root system, 7, 12, 13, 16, 17, 64, 91
Rothiemurcus Forest, 126, 137

Sampling, groundflora, 65, 68, 73, 113
—, litter, 12, 13
—, litter fall, 63
—, mesofauna, 79, 85, 98
—, roots, 17, 64
—, small mammals, 89, 91
—, trees, 60, 61, 68–72
—, woodland, 14–16
Saprophyte, 51, 53, **Ch. 7**
Saprovore, 54, 55, **Ch. 7**
Seasonal change, 18, 19, 58
Seed production, 11, 19, 32, 34, 42, 87, 89, 128
Seedling, tree, 10, 18, 20, 22, 30, 32, 35, 89
Senescence, tree, 20
Shade tolerance, 8, 20, 28, 110
Shifting cultivation, 42
Shrub, 9, 35, 44, 65, 68, 134–136
Similarity coefficient, 114
Soil, 26, 31, 53, 55, 58, 76, App. II
Species/area curve, **107**, 108, 113
Specific gravity, of stem, 102

Stand, 20
Stand cycle, 20, 22, 33
Succession, **Ch. 3**, 41, App. II
—, diverted, 30
—, inferred, 28
—, primary, 30
—, retrogressive, 28
—, secondary, 30, 34, 42
Suppression, of trees, 11, 20
Sweat Mere, 27–31, 34
Symbiosis, 95
Synusiae, 10
Systems ecology, 101

Thinning, forestry, 21, 61, 68
—, self, 21
Tolerance limits, 109, 110, 129
—, shade, 8, 20, 28, 110
Top Height, of stand, 72, 102
Transect, 14, 22, 73, 107, 108
Tree, dry weight determination, 68–7
—, longevity, 5

—, pioneer spp. 26, 35, 128–131
—, planting, 34, 46, 48, 89, App. II
—, seed production, 11, 19, 32, 34, 42, 87, 89, 128
—, seedlings, 10, 18, 20, 22, 89
—, senescence, 20
— suppression, 11, 20
Trophic level, 50, 59
Tullgren funnel, 80–83
Turnover, 11, 19, 28, 55, **Ch. 7**

Understorey, 9, 10

Woodland clearance, 32, 42, 45
— history, Ch. 4
— management, 1, 3, 21, 34, 46, 126
— sampling, 14–16
— types, Ch. 10
Wood rots, 86
Wytham Wood, Oxford, 91

Yield Tables, 72, 102, **103.**

Index to common and scientific names of species

Abies, 32, 39
Acari, 83, 84
Accipiter, 50
Acer platanoides, 89
A. pseudoplatanus, 85, 135
Actinomycetes, 76
Adelges, 132
Agrostis, 89
Alder, 25, 28, 35, 39, 42, 72, 128, 135
Algae, 11, 26
Alnus glutinosa, 25, 28, 35, 39, 42, 72, **128**, 135
A. incana, 128
Amoebae, 81
Anemone, 134
Apodemus, 91
Ash, 44, 85, **129**
Asio, 49

Bacteria, 54, 76, 95, 121, 132
Basidiomycetes, 6, 77
Beech, 19, 31, 32, 34, 42, 85, 112, 128, **131**
Beetle, 79, 81, 88
Bent, 89

Betula pendula, 41, 128
B. pubescens, 41, 128, 135
Bilberry, 65, 135, 137
Birch, 5, 34, 35, 39, 41, 128, 135
Black game, 88
Bluebell, 19, 65, 134
Bluetit, 50
Bracken, 10, 134
Bramble, 10, 134
Bupalus, 87, 98, 104–106
Buttercup, 134

Calluna, 137
Canis, 42, 93
Capreolus, 89
Carabidae, 79
Carex, 134, 135
Castanea, 128
Cat, wild, 93
Celandine, 49, 136
Cephalanthera, 131
Chamaecyparis, 136
Chermes, of Douglas fir, 132
Cherry, 133
Chestnut, sweet, 128

Clethrionomys, 91
Collembola, 83, 84, 94
Conifer, 37, App. II
Coral Spot, 85
Corylus, 134
Cotton grass, 138
Cowberry, 137
Crataegus, 35, 44
Crossbill, 97
Cypress, Lawson, 136

Daffodil, 136
Dama, 89
Deer, 42, 89, 91
—, fallow, 89
—, red, 91
—, roe, 89
Deschampsia, 135, 137
Diatrype, 85
Dog's mercury, 10, 131, 134

Earthworm, 80, 85
Elm, 41, 42, 133, 135
Enchytraeidae, 80, 85
Endymion, 19, 65, 134
Epipactis, 131
Eriophorum, 137

Fagus, 19, 31, 32, 34, 42, 85, 112, 128, 131
Felis, 93
Ferns, 25, 113
Ficaria, 49, 136
Fir, Douglas, 46, 132
—, silver, 32, 39
Fraxinus, 44, 85, 129
Fungi, 6, 53, 54, 74, 77, 78, 81, 85, 86, 95, 121

Grass, purple moor, 137
—, soft, 89, 134
—, wavy hair, 135, 137

Hawthorn, 35, 44
Hazel, 134, 136
Heather, 137
Hemlock, western, 132
Holcus, 89, 134
Holly, 135
Hylocomium, 137
Hypoxylon, 85

Ilex, 135

Juncus, 119, 134

Larch (*Larix* spp.), 10, 46, 132
—, European, 46, 132
—, Japanese, 132
Larix decidua, 46, 132
L. leptolepis, 132
Leaf miner, 88
Lepidoptera, 87, 98, 104–106
Lichen, 11, 26, 113
Lime, 42, 133
Liverwort, 9, 109, 113
Lophodermium, 74
Loxia, 97
Lyurus, 88

Maple, 40
Martes, 93
Mercurialis, 10, 131, 134
Mesostigmata, 82, 84, 94
Micro-arthropods, 79, 80, 82–84, 94
Microtus, 93
Millipede, 79
Mite, mesostigmatid, 82, 84, 94
Mite, oribatid, 54, 79, 83, 84
Molinia, 137
Moss, 9, 73, 113, 137
Moth, 87, 98, 104–106
Mouse, 91, 93, 94
Mustela, 93
Myelophilus, 88

Nanorchestes, 83
Narcissus, 136
Nectria, 85
Needle-cast, pine, 74
Nematode worm, 80
Neottia, 10, 94, 131
Nettle, stinging, 134

Oak, 5, 10, 31, 34, 39, 41, 42, 46, 85, 133–136
Oak, pedunculate, 41, 42, 133, 134, 136
Oak, sessile, 41, 112, 115, 116, 133, 135
Orchid, bird's nest, 10, 94, 131
—, Helleborine, 131
Oribati, 79, 83
Owl, long-earned, 49
—, tawny, 94
Oxalis, 135

Parus, 50, 94
Picea abies, 32, 39, 46, **137**
P. sitchensis, 46, 127, **137**
Pig, wild, 42
Pine, lodgepole, (beach, shore) 127, **129**
—, Scots, 5, 6, 13, 21, 31, 35, 39, 41, 61, 63, **136–7**
— marten, 93
Pinus contorta, 127, 129
P. sylvestris, see Scots Pine above
Pleurozium, 137
Poa, 25
Polecat, 93
Poplar, black, 34
Populus nigra, 34
Primrose, 19, 134, 135
Primula, 19, 134, 135
Protozoa, 54, 78, 81
Prunus, 133
Pseudotsuga, 46, **132**
Pteridium, 10, 134

Quercus petraea, 41, 112, 115, 116, 133, 135
Q. robur, 41, 112, 133, 134, 136
Q. spp. 5, 10, 31, 34, 39, 41, 42, 46, 85, **133–136**

Ranunculus, 134
Rhizobium, 95
Rhytidiadelphus, 137
Rowan, 135
Rubus, 10, 134
Rush, 119, 134

Salix, 35
Sanicula, 131, 134

Sedge, 134, 135
Sciurus, 88, 89, 97
Shoot borer, 88, 97
Sorbus, 135
Sparrowhawk, 50
Springtail, 83, 84, 94
Spruce, Norway, 32, 39, 46, **137**
Spruce, Sitka, 46, 127, **137**
Squirrel, 88, 89, 97
Stereum, 85
Stryx, 94
Sycamore, 85, 135

Teucrium, 135
Tilia cordata, 42, 133
T. platyphyllos, 42, 133
Titmice, 50, 94
Tortrix, 87
Trametes, 85
Tsuga, 10, 132

Ulmus, 41, 42, 133, 135
Urtica, 134

Vaccinium myrtillus, 65, 135, 137
V. vitis-idaea, 137
Vole, bank, 91, 94
Vole, short-tailed, 93

Willow, 35
Wolf, 42, 93
Wood-anemone, 134
Wood-sage, 135
Wood-sanicle, 132, 134, 135
Wood-sorrel, 135
Worm, 80, 85

Xylaria, 86